POLYMER-BASED SEPARATORS FOR LITHIUM-ION BATTERIES

POLYMER-BASED SEPARATORS FOR LITHIUM-ION BATTERIES

Production, Processing, and Properties

MARK T. DEMEUSE

ELSEVIER

Elsevier
Radarweg 29, PO Box 211, 1000 AE Amsterdam, Netherlands
The Boulevard, Langford Lane, Kidlington, Oxford OX5 1GB, United Kingdom
50 Hampshire Street, 5th Floor, Cambridge, MA 02139, United States

Library of Congress Cataloging-in-Publication Data
A catalog record for this book is available from the Library of Congress

British Library Cataloguing-in-Publication Data
A catalogue record for this book is available from the British Library

ISBN: 978-0-12-820120-6

Publisher: Matthew Deans
Acquisitions Editor: Edward Payne
Editorial Project Manager: Ruby Smith
Production Project Manager: Nirmala Arumugam
Cover Designer: Victoria Pearson

Typeset by TNQ Technologies

Contents

Preface

Living in today's world means that it is virtually impossible to go a day without coming into contact with a lithium-ion battery. These batteries are currently powering cellphones, smart watches, tablets, laptops, cars, bikes, homes, and planes. The process of selecting cells and the details of manufacturing are important to guaranteeing the adequate performance of the lithium-ion battery.

In addition, lithium-ion batteries are increasingly being used in emerging technologies such as electric vehicles and power grid applications. Lithium-ion batteries show promise for creating smaller, lighter, and cheaper devices for high-end applications. These applications enable these batteries to be used to develop consumer electronic cellphones and laptop computers.

A key component in the lithium-ion battery is the separator. It provides a barrier between electrodes while allowing the exchange of ionic charge carriers from one side of the battery to the other. As such, the separator performs two important functions vital to the performance of the battery.

The properties of the battery separator can be varied by using different processing scenarios. A complete understanding of the structure—property—processing relation for separator materials is necessary to optimize battery performance. Also, the effect of different separator structures on other components in the lithium-ion battery needs to be considered. A theme of this work is that the battery should be regarded in its entirety and not as individual materials that do not affect each other.

This book begins by explaining polymeric materials that can be used for separators, and characterization techniques for the separator materials. After that, available technologies used to produce separators for lithium-ion batteries are presented. Each technology is discussed in terms of advantages and disadvantages of the particular approach. Then, the properties of separators that are made by each technology are summarized and compared. Finally, areas for further development are addressed and the limitations of current materials and separators to obtain those goals are highlighted.

All current and new technologies used to produce polymer battery separators for lithium-ion batteries are covered. An analysis is provided for connections among different materials and processes. Advantages and disadvantages and the resulting properties from each technology are discussed. The work that is presented will enable the reader to develop polymer separators that meet requirements in terms of properties, performance, and industry standards. It will allow the reader to select materials and processes to develop high-performance polymer separators with enhanced properties.

The book will serve as a valuable resource for industrial scientists working on polymer-based separators as well as new materials and processes for lithium-ion batteries. In the academic world, the work will be of interest to researchers and advanced scientists in several different disciplines, including polymer science, materials science, energy, and chemical engineering. They should all benefit from the information provided here.

Whenever a project of this magnitude is undertaken, there will be many people who will have provided assistance to the author. There are numerous individuals whose help should be recognized. First, all of my technical colleagues who have provided challenging discussions are noted. Second, all of my consulting clients who have allowed me to explore various polymeric separators are thanked. Finally, all of the people at Elsevier who have assisted with the production of this book are acknowledged for all of their valuable contributions.

Mark T. DeMeuse
MTD Polymer Consulting
Charlotte, NC, United States

CHAPTER 1

Introduction to lithium-ion battery design

Introduction

A lithium-ion (Li-ion) battery (LiB) is a type of rechargeable battery. Sony and Asahi Kasei released the first commercial LiB in 1991. The global LiB market was valued at $36.7 billion in 2019 and is projected to reach $129.3 billion at a compound annual growth rate of 18.0% from 2020 to 2027. LiBs have major applications in laptops, cell phones and iPods.

Lithium batteries were proposed by M. Stanley Whittingham while working for Exxon in the 1970s. The initial batteries used titanium disulfide and lithium metal as the electrode. This rechargeable battery could never be made practical. Titanium disulfide had several issues, including its high cost. In addition, batteries with metallic lithium electrodes caused safety issues. Research thus moved to develop batteries in which only lithium compounds are present and were capable of accepting and releasing lithium ions.

In the 1980s, John B. Goodenough experimented using lithium cobalt oxide (LCO) as the cathode instead of titanium disulfide. This allowed the battery to double its energy potential. Five years later, Akino Yoshino used petroleum coke instead of reactive lithium metal as the anode. This led to a new battery that was not only significantly safer without the use of lithium, the battery performance was more stable. Thus, the first prototype of the LiB was produced.

Together, these three discoveries led to the LiB as we now know it. The initial introduction of LiBs was driven by the need for a lightweight rechargeable cell to power the rapidly growing market for portable electronic equipment in the 1990s. Since then, these batteries have been employed in many consumer applications as well as military and aerospace uses. They provide high-energy capacity compared with other types of batteries.

Polymer-Based Separators for Lithium-Ion Batteries
ISBN 978-0-12-820120-6
https://doi.org/10.1016/B978-0-12-820120-6.00002-7

Today's world of portable electronics means that it is virtually impossible to go a day without being in contact with an LiB. These types of batteries power up cell phones, smart watches, tablets, laptops, cars, bikes, homes, and planes. The cell selection, manufacturing process, and electrical and mechanical design of the battery are key components to ensure the adequate performance of the battery.

LiBs are also used in emerging technologies such as hybrid electric vehicles, which use batteries as well as electric motor engines to increase fuel efficiency [1]. The electrical energy that a battery provides is a function of both the cell and its capacity, which depend on the chemistry of the battery. LiBs show promising advantages in creating smaller, lighter, and cheaper batteries for high-end applications [2]. Primarily because of these factors, these batteries are also widely employed in developing consumer electronics for new cell phones and laptop computers [3].

Fig. 1.1 shows a typical LiB construction. It consists of three major components that make up the battery: cells, housing, and electronics. The cell is the power source of the battery. It comes in many different sizes, shapes, and chemistries. The primary purpose of the electronics is to guarantee that the cells are used within safe operating conditions. The housing provides an enclosure for the cells and board for mechanical integrity.

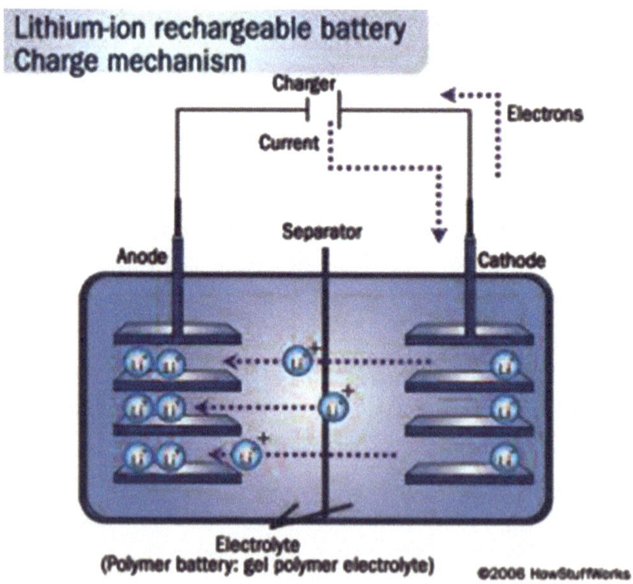

Figure 1.1 Typical view of lithium-ion battery construction.

Lithium-ion cells

The use of LiBs has grown significantly in recent years. They offer some distinct advantages and improvements over other forms of battery technology. However, like all technologies, LiBs also have disadvantages.

There are many advantages to using an LiB. Among these are high-energy density, self-discharge, low maintenance, cell voltage, load characteristics, no requirement for priming, and the variety of types that are available. These advantages mean that these batteries find an increasing number of applications; as a result, a huge amount of development is being invested into them.

However, there are certain disadvantages to using LiBs. Among these are that they are relatively expensive to manufacture and they show sensitivity to high temperatures. Additional concerns are the aging effect the batteries experience, deep discharge that makes the battery unusable, and general safety concerns associated with their operation and use.

Li-ion cells come in three main forms: cylindrical, prismatic, and polymer.

A cylindrical cell comes in two common sizes: 18650 and 26650. The cells are either steel or aluminum. The first two digits in the size is the diameter of the can in millimeters, and the next three numbers indicate the length in millimeters. So, for an 18650 cell, shown in Fig. 1.2, the diameter is 18 mm and the length is 65.0 mm. The 18650 cell is one of the most popular cell packages. Typical applications for the 18650 Li-ion cells are power tools, medical devices, laptops, and e-bikes.

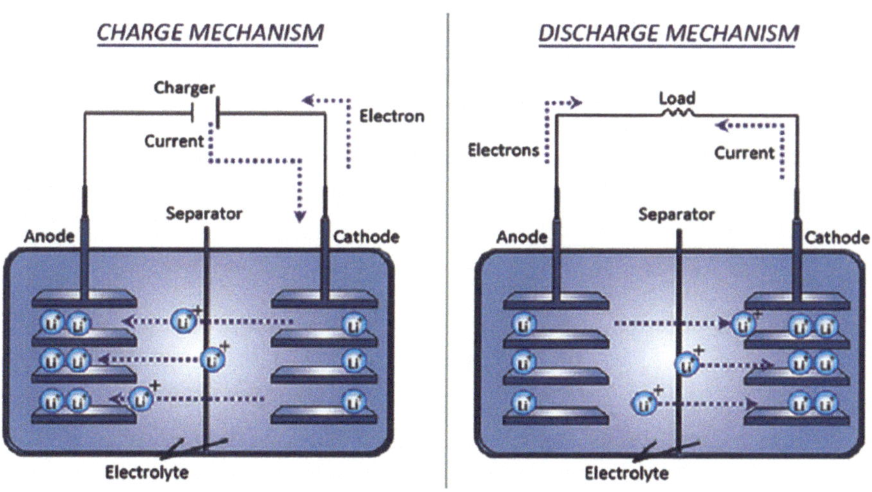

Figure 1.2 Li-ion battery during discharge and charge.

Prismatic cells satisfy the demand for thinner sizes. They make optimal use of space by using the layered approach. Other designs are wound and flattened into a prismatic-type jelly roll. These cells are found in mobile phones, tablets, and low profile laptops. No universal format exists and each manufacturer designs its own.

An Li-ion polymer battery differs from these two other types only in construction. It does not have a unique and different chemistry. The difference in construction allows for lower cost, safer operation, and more flexible packaging options. Li-ion polymer battery packs can be as thin as a credit card. These factors make them an increasingly popular choice for a wide range of applications.

The choice of which battery cell design to use depends on several factors including space, cost, and safety considerations. The cylindrical cell design has good cycling ability, offers a long calendar life, and is economical. On the other hand, it is heavy and has a low packaging density owing to cavities of space.

The prismatic cell's key advantages lie in its thin profile, lightness, and effective use of space. The prismatic cell's thin, rectangular shape allows for effective layering and provides designers with increased flexibility. Prismatic cells are typically found in mobile phones, tablets, and other lightweight electronic devices.

Prismatic cells are expensive to design and manufacture. This makes them more expensive for consumers. They have shorter lifetimes because thermal management is less effective. In general, they are relatively sensitive to deformation in high-pressure situations. Finally, prismatic cells are available in a limited number of standard sizes and have an average higher watt per hour price than cylindrical cells.

Currently, the most effective cylindrical lithium cell controls market share. However, as new products create an increasing need for space-conscious solutions, demand for prismatic cells is expected to increase. Prismatic cells are the best device for applications that require power packed in a small space.

Cell components

In each of the next four sections of this chapter, descriptions of the cathodes, anodes, separators, and electrolytes will be provided. Every single component of the LiB is essential because it cannot function when one of the components is missing. Because the focus of this book is the battery separator, descriptions of each material will be relatively brief and will focus on the function of the various battery components. More details of the separator will be provided in subsequent chapters.

Cathodes

Cathodes have always been thought of as the primary factor related to the high capacity and superior power delivery of LiBs. The cathode undergoes reduction, or a gain of electrons, during the discharge process. The lithium ions accumulate in the cathode. Upon charging, on the other hand, the movement of the electrons is reversed and the cathode regains the higher potential. The cathode remains at a higher potential during discharging of the battery.

Traditionally, lithium metal oxides have been used as the cathode material in LiBs. It is generally accepted that having a cathode capable of enhancing battery capacity has a negative effect on its life cycle. In general, a balance between the capacity and the life cycle of the battery will lead to more efficient batteries.

In general, the choice of cathode material with a particular chemistry depends on various factors including cell voltage, capacity, energy and power capabilities, life cycle, and temperature of operation. LCO is one of the most widely used materials for cathodes. However, there are cost, safety, and environmental hazards associated with LCO. These concerns have led to the development of various other cathode materials. In general, the type of cathode material that is chosen can affect the energy density, power density, safety, life cycle, and cost of the overall battery. That is why researchers are constantly experimenting with new ideas and material combinations.

For materials to be used as cathodes in LiBs, Li ions must be able to move freely through the crystal structure. The structure and morphology of the crystal, which is one-, two-, or three-dimensional, largely determines the number of dimensions in which the lithium ions are able to move. Three primary morphologies for cathode materials are currently used or are under development.

First, there are olivine structure materials. Of these, lithium iron phosphate is the most well-known. These structures restrict the diffusion of lithium ions in a single linear dimension. Although low ion mobility poses a performance disadvantage coupled with low discharge voltage, there has been commercialization of olivine cathode materials because of the outstanding stability that they provide.

Next, there are layered rock salt structure materials. This group of compounds has a two-dimensional crystal morphology. Lithium nickel oxide and lithium manganese dioxide are two examples of this class of cathode materials. Generally, they are combined with other elements to form complex oxides such as lithium nickel manganese cobalt oxide. These batteries can have either high specific energy or high specific power. They also have a high number of charge and discharge cycles. These batteries are most common in power tools and powertrains for vehicles.

Nickel, cobalt, and aluminum oxide (NCA)-LiBs use a combination of 80% nickel, 15% cobalt, and 5% aluminum. The addition of the aluminum to the NCA battery makes it more stable. Batteries that use NCA are often employed for high-energy applications such as backup power by utility companies. These types of batteries have about 60—70 charge and discharge cycles per year with a total of about 1500 cycles.

There are also spinel structure materials that exhibit a three-dimensional crystal structure. Of these materials, lithium manganese oxide is the most significant compound. Spinels offer advantages in terms of lower cost and high stability and therefore are gaining attention in emerging applications for LiBs. However, they have lower discharge capacity than two-dimensional materials.

Anodes

The anode is the part of the battery from which electrons move toward the cathode during a discharge process and from the cathode upon a charging operation. The anode, in conjunction with the cathode, produces the required output needed for the battery to operate. To be suitable for LiB manufacturing, anode materials should meet following requirements of:

- Excellent porosity and conductivity
- Good durability and lightweight
- Low cost
- Voltage match with preferred cathode material

Primarily because of its high specific energy per weight, metallic lithium was initially proposed as the material of choice for anodes in LiBs. However, lithium metal has high instability and the growth of lithium dendrites leads to disastrous results. Safety issues shifted the focus to safer anode materials such as coke and graphite. These safer materials have a disadvantage compared with lithium metal involving lesser specific energy capacity. The use of graphite as the anode material has surpassed that of any other material primarily because of its ability to yield a flat discharge curve. Research in the field is ongoing to produce optimized anode materials.

Graphite remains one of the most commercially attractive anode materials for use in LiBs. Graphite anodes meet the voltage requirements of most common Li-ion cathodes; they are relatively affordable, extremely light, porous, and durable. Another designation for electrochemically active graphite is meso-carbon microbead. Because there is little scope to enhance the capacity of a graphite anode further, additional research has largely focused on other new materials such as metal oxides that include cobalt oxide, copper oxide and lithium–metal alloys.

The use of blends or mixtures allows for maximizing the amount of lithium at the anode upon discharge. Blending permits increased capacity to be achieved compared with graphite alone while maintaining good conductivity and a low electrochemical potential. The use of blends increases the maximum amount of lithium that can be intercalated within the blended anode structure. The performance, potential, and availability of each blended material are all preferable when a comparison is made with current graphite-based anodes. The geometry of the blended material can be engineered to yield no significant swelling or stack pressure generated by the anode upon prolonged cycling. Therefore, Li-ion cells can be constructed as cylindrical, flat, or prismatic cells with thin-walled cases or in any other cell configuration.

Crystalline silicon or tin anodes have a theoretical specific capacity more than ten times that of anodes made from graphite. However, expansion causes large anisotropic stresses to occur within the electrode material. Those stresses lead to fracturing and crumbling of the material and detachment from the current collector. Lithium-silicon batteries lose most of their capacity in as little as 10 charge–discharge cycles. A solution to capacity and stability issues caused by volume expansion is critical to the success of silicon and tin anodes. Nanowire morphology can overcome problems that are associated with the stability of these anodes. This is done by providing conductive pathways for the ions and by decreasing the disruption in volume change. In addition, nanoparticles with the use of polymeric binders are being employed to achieve similar effects to deal with costs incurred by the use of nanowires.

Separators

Because separators are the main topic of this book, the discussion that is presented here will be brief and will provide only general details on the topic. Additional and more specific information will be provided in subsequent chapters.

A good review of separator materials and needs was provided by P. Arora and Z. Zhang [4]. As the name implies, the battery separator separates the two electrodes physically from each other. Batteries use separators to prevent short-circuiting of the electrodes. Microporous polymer-based separators are used to block physical contact between the anode and the cathode. Polyethylene and polypropylene membranes are used as separators. In addition, many polymer separators, typically multilayer polymer separators, act as shutdown separators able to shut the battery down if it becomes too hot during the cycling process. These multilayer polymer separators are typically composed of one or more polyethylene layers that function to shut the battery down, and at least one polypropylene layer that acts to provide mechanical support for the separator.

The desired features of a separator include its thickness uniformity, adequate mechanical strength during cell fabrication, and chemical and electrochemical stability when in contact with the electrolyte and during the charge–discharge cycle. Separators must be chemically stable when they contact electrolyte and electrode materials that are being used. They should not undergo degradation during the process of charge and discharge. The separator affects the internal cell resistance, stability, life cycle, operating temperature range, and cell kinetic parameters such as the discharge and charge rates. Through its ability to regulate electrolyte distribution between the active electrodes, as a result of its morphology and resultant permeability features, the separator limits ionic diffusion and recombination rates. Its presence thus affects the cell capacity, power, and available energy. With respect to the inherent safety of an LiB, the separator can provide passive protection and tolerance to cell abuse conditions that may arise.

In LiBs, dendritic formations are unavoidable; this leads to the phenomenon of growing electrodes. When the dendrites from the electrodes meet each other, the electrodes essentially become short-circuited. To avoid this situation, batteries use a separator that has small openings or pores to allow for the diffusion of ions. The separator acts as a barrier for dendrites and hinders growth in unwanted directions. The downside of using a separator is the hindrance that it creates for the diffusion of ions between electrodes. This reduces the response time and increases the charging time. Efforts are currently under way to use solid-state systems such as solid electrolytes to produce more efficient batteries.

Although not necessarily an active component in a cell, the separator has a key role in ion transport and influences the rate performance, cell life, and safety. However, it is becoming evident that there are opportunities to improve separators to meet the greater demands that new applications place on LiB technology. LiB performance can be improved by considering the interplay of separators with their surroundings. Indications are that in the future, separators will be designed to have a more active role in LiB operation. Current and developing characterization techniques will have a significant role in guiding this evolution in separator technology.

Thus far, all commercial separators have been made of polyolefins, but they provide only limited heat resistance. Research now focuses on separators made of different materials that could offer superior heat resistance. Separators made of such materials are expected to show high temperature stability and safety as well as superior ion transport for better rate capability at high current discharge. The overall stability of the separators, under potentially extreme battery operating conditions, will require leading edge design and fabrication techniques to exceed manufacturing and performance requirements for large-scale applications, such as electric vehicles. Good thermal stability has been a primary focus of development efforts in this regard. A summary of these research efforts into different materials will be provided in subsequent chapters of this book.

Electrolytes

High-purity electrolytes are a core component of LiBs. The electrolyte has a key role in transporting lithium ions between the cathode and anode. The electrolyte should completely wet both the separator and the electrodes to reduce cell internal resistance. Current commercial LiBs use liquid organic electrolytes, which have significant advantages of high conductivity and excellent wettability on electrode surfaces. The most commonly used electrolyte is composed of a lithium salt such as LiPF6 in an organic solution. Common solvents are a mixture of carbonate esters such as ethylene carbonate and propylene carbonate and other esters such as dimethyl carbonate and diethyl carbonate.

Electrolyte solutions need to enable lithium ions to move freely. This requires a high dielectric constant as well as low viscosity. Suitable electrolytes are a combination of two esters, as previously described. Ideally, the electrolyte solution will have a wide range of voltage stability with adequate Li-ion conductivity.

Electrolytes are often a solution of lithium salts, usually LiPF6, in a stable organic solvent. The lithium salt remains in the form of dissociated ions in the solvent, allowing for the movement of these ions. The diffusion of ions helps maintain the electrical neutrality of the overall battery. Major issues associated with current electrolytes are their high flammability and slow diffusion. These concerns are being addressed through the use of electrolytes based on solid-state materials with inherent low flammability.

Research on electrolytes generally focuses on one of three areas: functional electrolyte additives, flame resistant or nonflammable electrolyte solutions, and new electrolyte solutions. A brief discussion of each of these areas of research will be provided. First, the area of functional electrolyte additives will be discussed. A review of this topic was provided by Zhao et al. [5]. That summary gives a short overview of work on some functional electrolyte additives.

In general, they are one of the key materials that affect the electrochemical performance of LiBs, such as energy and power density and battery cycling performance. An early example of this approach is the addition of propane sulfone to a nonaqueous electrolyte solution. A range of compounds were used as additives; important examples include vinylene carbonate, propane sulfone, phenyl-cyclohexane, and fluoroethylene carbonate. The choice of appropriate additives and the specification of required formulations have become an important aspect of proprietary knowhow for each battery manufacturer.

The main approach to flame-resistant electrolyte solutions is to employ phosphate compounds in some manner, by using cyclic phosphoric acid ester as a solvent [6] or by adding a phosphazene compound as a retardant material [7]. The next most common approach is to use halogen compounds, particularly fluorine compounds, such as fluorocarbon ester and fluorinated ether, as the solvent [8]. A further idea is a solvent mechanism that involves encasing a flame retardant in microcapsules released in the case of a battery malfunction [9].

Another area of research is the use of electrolyte salts to replace LiPF6. This approach provides numerous challenges in terms of performance and cost. A review is available of various salts that can be used to conduct Li^+ ions in electrolyte solutions [10]. In that work, several types of lithium salts, such as bis(oxalate)borate, oxalydiflurorοborate, and fluoroalklyphosphate, are described in detail. The prospect of using these lithium salts in LiB is described.

Finally, solid-state electrolytes need to be discussed. The primary motivation for using this type of material is safety issues associated with flammable organic electrolytes. There is also the potential for improving the conductivity of lithium ions through solid-state electrolytes that could even surpass traditional liquid electrolytes [11]. An excellent review of progress in this area was provided [12]. That review summarized a series of all solid-state polymer electrolytes based on modified poly(ethylene oxide), polycarbonate, polysiloxane, succinonitrile, and organic—inorganic hybrid composites. The authors concluded that that high-performance solid-state polymer electrolytes were expected to be used in portable electrochemical devices, electric vehicles, and grid energy storage. However, the use of solid-state electrolytes remains in the development stage, and further work appears to be necessary to advance the efforts to the stage of full commercialization.

Returning for a moment to the case of liquid electrolyte materials, the decomposition materials of the electrolyte serve an important role in the overall performance of the battery. The solid electrolyte interphase (SEI) is a layer that forms on the anode surface for all alkali metal batteries that use liquid electrolytes [13]. The construction of a stable and efficient SEI is among the most effective strategies for inhibiting dendrite growth and achieving superior cycling performance in the battery. In fact, to have a healthy and high-performing LiB, the quality of the SEI layer is critical. During early cycling of the battery, the SEI layer mitigates initial capacity loss, which improves the overall battery safety and a cell's lifetime. A uniform SEI layer protects and electronically insulates the anode surface. At voltages greater than 4.2 V, this will prevent degradation of the electrolyte during battery cycling.

Ideally, the SEI layer offers sufficient electronic resistance to limit electrolyte decomposition to the amount needed for its formation. However, slow and continued SEI growth lead to capacity fade and increased cell resistance in the battery [14,15]. During the long-term cycling of LiBs, the formed SEI layers either grow in thickness or become compromised at the last stage.

Several methods have been proposed that focus on inhibiting the detrimental chemical reaction in the SEI layer to improve the interphasal stability between the electrode and the liquid electrolyte. Among those approaches, there are two main developments. First, electrolyte additives such as vinylene carbonate [16] and fluoroethylene carbonate [17—19] have been used to suppress unwanted decomposition of the electrolyte. In general, use of these types of electrolyte additives is one of the most effective and economical ways to improve the cyclability of LiBs. They also increase the discharge capacity retention of the batteries.

The other approach involves the development of solvents such as nitriles [20,21], sulfoxides [22], and ionic liquids [23] to replace conventional carbonate solvents to modify the properties of the SEI layer. These types of effects have led researchers to correlate the structure and property of each SEI.

There are two types of components in the SEI layer: inorganic components, such as LiF, Li_2O, $Li_2PO_3F_2$, Li_2PFy, and Li_2CO_3, and organic materials such as $ROCO_2Li$, $(CH_2OCO_2Li)_2$, and ROLi [24,25]. The contributions of each of these materials to the cycling performance of the batteries has been investigated and their mechanical and transport properties have been investigated by simulations [26]. The material features of each component have been related to the performance of the lithium - ion batteries.

Among the components described, LixPOyFz has often been discussed. For example, Li et al. [27] found that the SEI layer was covered by more LixPFyOz when prop-1-ene-1,3-sulfone was used as an electrolyte additive. It was thought that a protective SEI layer usually contains LixPOyFz. That layer inhibits side reactions between the electrolyte and the electrode, improving the cyclability performance of the LiBs.

In a further study, Yang et al. [28] incorporated $LiPO_2F_2$ as an electrolyte additive into the reference electrolyte. Impedance spectroscopy indicated that the SEI layer containing the $LiPO_2F_2$ could decrease the surface impedance of cells in the last cycle stage. The $LiPO_2F_2$ is deposited on the surfaces of both the $LiCoO_2$ and graphite electrodes. This has the effect of protecting the graphite anode and suppressing degradation of the cathode during the long-term cycling of the batteries.

A review was provided of state-of-the-art modeling progress in investigating the development of the SEI layer [29]. That review concluded that the ideal SEI layer should possess two features: electrically insulating and Li-ion conducting. Furthermore, various modeling methods such as electronic, atomic scale, and phenomenological models are deemed appropriate to provide insight into the SEI layer. Topics covered in that review include the thermodynamics and kinetics of electrolyte reduction reactions, SEI formation, modification through electrolyte design, and the correlation of SEI properties with battery performance. Multiscale simulations are summarized and compared with each other as well as with experiments. The review concludes that computational modeling can be integrated with experiments to complement each other and lead to a better understanding of the SEI layer.

Cell chemistry

The overall construction of LiBs is similar regardless of the actual chemistry used. The battery consists of four main parts [1]: anode [2] cathode [3], separator, and [4] electrolyte. The overall goal of the battery is to create a flow of electrons that creates a current that can be used to power various devices.

The electrodes consist of a cathode and anode and are made of two dissimilar materials. One electrode is free to provide electrons whereas the other is more willing to accept electrons. The separator guarantees that the anode and cathode do not touch and cause a short circuit. At the same time, the separator allows Li-ion flow to flow through. Separators are the topic of this book and will be described in more detail in other chapters. The electrolyte in the LiB is used to create the oxidation–reduction process and serves as a medium for lithium ions to flow between the anode and cathode.

When a circuit is completed, as shown in Fig. 1.2, a chemical reaction is created that allows for the discharge and charge of the battery. The anode material provides electrons on the negative side; they flow to the cathode side of the battery. At the same time, lithium ions flow from the cathode to the anode through the electrolyte material. The opposite effect happens during the charging process.

Generally, in lithium - ion batteries, the anode material is carbon or graphite. Various materials have been used as the cathode material. A list of common lithium chemistries is provided in Table 1.1.

Table 1.1 Common lithium chemistries.

	Lithium cobalt oxide	Lithium iron phosphate	Lithium manganese oxide	Lithium nickel Manganese cobalt Oxide	Lithium nickel Cobalt aluminum Oxide
Anode	Graphite	Graphite	Graphite	Graphite	Graphite
Cathode	Lithium cobalt oxide	Lithium iron phosphate	Lithium manganese oxide	Lithium nickel Manganese cobalt oxide	Lithium nickel cobalt Aluminum oxide
Chemistry	$LiCoO_2$	$LiFePO_4$	$LiMn_2O_4$	$LiNiMnCoO_2$	$LiNiCoAlO_2$
Name	LCO	LFP	LMO	NMC	NCA
Nominal voltage	3.6 V	3.2 V	3.7 V	3.6 V	3.6 V

Table 1.1 compares the most common Li-ion chemistries available on the market.

In terms of energy density, Li-ion chemistry provides twice that of standard nickel-cadmium batteries. A big advantage is the battery pack design, which requires the use of only one cell. This is possible owing to the cell voltage of 3.6 V. On the other hand, nickel-based batteries require the use of three 1.2-V cells connected in series. Most mobile phones run on a single cell.

Another big advantage of Li-ion chemistry is low maintenance. In addition, because there is no battery memory, there is no need for scheduled cycling to prolong the life of the battery. Furthermore, the self-discharge of an LiB is less than half compared with that of a nickel-cadmium battery.

Although Li-ion chemistry has many advantages over other possible cell chemistries, the biggest drawback is the safety performance. An Li-ion cell can create a thermal event if not used properly. Li-ion is a volatile chemistry that has many safety concerns over other battery chemistries. However, the high-energy density, high life cycle, and maintenance-free operation make it attractive to use Li-ion cells to power electronics. Safety concerns can be addressed by selecting appropriate lithium cells for the desired application. Also, the proper electrical and mechanical design to manage the cells must be used.

Battery operation

The reactants in the electrochemical reactions in an LiB are the negative and positive electrodes and the electrolyte. The electrolyte provides a conductive medium for the lithium ions to move between the electrodes. Electrical energy flows out from or into the battery when electrons flow through an electrical circuit during either discharge or charge.

Both of the electrodes allow for the movement of lithium ions in and out of their respective structures by a process called intercalation or deintercalation, respectively. During the discharge process, the positive lithium ions move from the anode to the positive cathode. The ions move through the electrolyte while the electrons flow through the external circuit in the same direction. On the other hand, when the battery is charging, the reverse process occurs and the lithium ions and electrons move back into the negative electrode.

The positive electrode half-reaction in the lithium containing cobalt oxide substrate is [30,31]:

$$CoO_2 + Li^+ + e^- \rightarrow LiCoO_2 \tag{1.1}$$

The negative electrode half reaction for the graphite is:

$$LiC_6 \rightarrow C_6 + Li^+ + e^- \tag{1.2}$$

Thus, the full reaction is:

$$LiC_6 + CoO_2 \rightarrow C_6 + LiCoO_2 \tag{1.3}$$

The overall reaction has its limitations. Overdischarge of the battery supersaturates lithium cobalt oxide, leading to the production of lithium oxide [32]. This likely occurs through the reaction:

$$Li + e^- + LiCoO_2 \rightarrow Li_2O + CoO \tag{1.4}$$

Trends in lithium-ion batteries

With the growing demand for electric vehicles and smaller, faster electronic devices, the Li-ion battery industry is constantly evolving. The need for safer, lighter, and longer-lasting batteries is more pertinent than ever before, and those involved in the manufacturing and formulation of Li-ion components continue to develop innovative solutions that satisfy industry requirements.

Li-B chemistry is comprehensively competing with other traditional battery types such as lead-acid batteries. LiB material industry growth is being driven primarily by end-user industries such as automotive and consumer electronics. It is witnessing a high growth rate and is expected to show a similar trend over coming years.

With growing emphasis on electric vehicles, demands on material suppliers are changing. With increasing demand from the electric vehicle sector, the production of LiBs is increasing rapidly. Specifically, this means that the LiB separator market has witnessed positive growth in recent years.

Favorable government initiatives that promote the use of electric vehicles as well as rising levels of emissions are two major factors driving the growth of the LiB market. Shifting preferences toward sustainable sources of energy in the power generation sector further aid market expansion. In addition, energy storage systems provide both economical and sustainable solutions to meet the growing demand for electricity. Increasing investments for developing large-scale energy storage systems is also stimulating industry growth.

A single electric vehicle car uses about 6000 times more separator than does a cell phone battery. With growth in the electric vehicle industry, the leading driver for the LiB separator market has shifted from consumer electronics such as laptops and cell phones to electric vehicles. Also, other material developments are occurring with electric vehicles that involve electrode materials. It is not completely clear that the current separators are optimal ones to use with these anodes and cathodes. These developments require the LiB to be examined in its entirety rather than its individual components.

This is a relatively new way to think about components in an LiB. In the past, many developments of materials for these batteries have been done in isolation. For example, anode materials have been developed without considering how these materials could affect the performance of the separators and the electrolytes. This approach is changing, however, and a more holistic way of developing LiBs is being pursued. This has necessitated more of a multidisciplinary outlook to future developments.

This holistic approach has been applied to screening electrolyte materials for use in LiBs [33]. In that study, it was realized that electrolyte materials must satisfy many requirements at once. Such an optimization is difficult to perform experimentally. With that realization, the authors develop a logistic regression approach to identify candidate structures likely to show fast lithium conduction. The approach is able to reduce the number of candidate materials significantly and has identified new materials, few of which have been examined experimentally.

A main driving force for further materials developments for LiBs is the emergence of applications such as electric vehicles, power tools, and energy grid storage systems. A factor that distinguishes portable electronics from electric vehicle applications is power density. The latter application requires a much higher charge and discharge rates than the former. The use of nanomaterials is expected to have a key role in achieving this high power capability. Nanomaterials are also employed in high-capacity electrodes so that the storage capacity can be used reversibly with a minimal compromise in the battery life cycle. Overall, the use of nanomaterials in electrodes will provide a new generation of electrical energy storage options for smart grids and backup power systems.

In addition to nanomaterials as electrode materials, silicon is attracting attention as an anode material in LiBs [34]. This is primarily because of the high specific capacity that silicon provides. However, the poor life cycle, irreversible capacity loss, and issues with the SEI layer [35] have prevented more practical implementation of silicon anodes. Nevertheless, Wu et al. [36] reported silicon-based composite electrodes with relatively stable reversible capacity and stable performance. Further research is needed to address the practical requirements of silicon anodes, such as high power density and low cost.

Graphene-enhanced silicon anodes have been developed to deal with providing hundreds of charge—discharge cycles. In this case, the high-capacity silicon materials are combined with mechanically reinforcing and electrically conductive graphene. The capacity of the material can be tailored to the specific application by mixing it with other conductive additives.

With the growing demand for electric vehicles and smaller, faster electronic devices, the LiB industry continues to evolve. The need for safer, lighter, and longer-lasting batteries is more relevant than ever before. Those involved in manufacturing and formulating the battery components continue to develop innovative solutions that satisfy industry requirements. The need for longer-lasting, higher-performing materials continues to increase. The expected performance necessitates higher voltages, increased stability, and improved capacity retention. These needs translate into more robust cell components in the electrolyte, separator, and SEI layer formulations.

Conclusions

This chapter has provided background information on the overall design of LiBs. It has discussed, in a general way, the various components in an LiB and the function of each of those components. Also, it was stressed that developments in these batteries will require more of a holistic approach than was previously used. This is different from past approaches in which each components was in relative isolation from each other. This means that the various parts of the final battery have not been optimized for use with each other.

In addition, new materials developments in battery components other than the separators were briefly reviewed. Because separators are the topic of this book, developments for them will be discussed in the remainder of the chapters. Other material developments should be considered in their relation to the separators used in the batteries. This will allow for the optimum overall battery system to be defined for new and emerging applications.

Most current developments in the design of LiBs are being driven by the electric vehicle market. In general, that market is focused on batteries of a larger size, and which have specifications different from those for batteries used in consumer electronics applications. This means that the materials need to be examined in a different way and with different features. However, if a new application emerges, there is no guarantee that the materials that have been defined as optimum for use in electric vehicles will be best for this new use. That is why it is critical for work to continue on a general understanding of the structure of one part of the battery as it relates to the other components.

The LIB has clear advantages, and decades of research have developed it into the high-energy density, high-life cycle, high-efficiency battery that it is today. Yet, research continues on new battery component materials to push the boundaries of cost, energy density, power density, life cycle. and safety. As new materials are found, LiBs will no doubt have an even greater impact on the lives of people in years to come.

LiBs are likely to remain commercially important for a vast array of energy storage applications for a long time. This means that making further improvements to their current operational limitations continues to be valuable. Only through the development and implementation of such improvements, can the full use of LiBs in such applications be completely realized.

References

[1] K.A. Smith, IEEEE 30 (2) (2010) 18–25.
[2] O. Gross, J. Sawyer, IEEE Battery Conf. Appl. Adv. 1 (2002).
[3] J.M. Tauscon, M. Armand, Nature 414 (6861) (2001) 359–367.
[4] P. Arora, Z. Zhang, Chem. Rev. 104 (10) (2004) 4419–4462.
[5] W. Zhao, Y. Ji, Z. Zhang, M. Lin, Z. Wu, X. Zheng, Q. Li abd, Y. Yang, Curr. Opin. Electrochem. 6 (1) (2017) 84–96.
[6] T. Shiga, C. Okuda, Y. Kato, H. Kondo, J. Phys. Chem. C 122 (18) (2018) 9738–9745.
[7] X. Li, W. Li, L. Chen, J. Power Sources 378 (2018) 707–716.
[8] G. Yan, X. Li, Z. Wang, H. Guo, W. Peng, Q. Hu, J. Wang, J. Solid State Electrochem. 21 (6) (2017) 1589–1597.
[9] M. Baginska, N. Sottos, S. White, ACS Omega 3 (2) (2018) 1609.
[10] P. Huang, S.J. Chang, C.C. Li, J. Power Sources 388 (2017) 82–90.

[11] V. Aravindan, J. Gnanaraj, S. Madhawi, H.K. Liu, Chemistry 17 (31) (2011) 14326–14336.
[12] Y. Manthiram, Wang, Nat. Rev. Mater. 2 (2017) 16103.
[13] L. Yue, J. Ma, J. Zhang, J. Zhau, S. Dong, Z. Liu, G. Cui, L. Chen, Energy Storage Mater. 5 (2016) 139.
[14] Y. Mo, S.P. Ong, G. Ceder, Chem. Mater. 24 (2010) 15.
[15] C. Bonmier, X. Ji, Small 14 (2018) 1703576.
[16] T. Waldman, M. Wilka, M. Kasper, M. Fleischhammer, M. Wohlfahrt - Mehrens, J. Power Sources 262 (2014) 129–135.
[17] J. Vetter, et al., J. Power Sources 147 (2005) 269–281.
[18] D. Aurbach, K. Gamolsky, B. Markovsky, Y. Gofer, M. Schmidt, U. Heider, Electrochim. Acta 47 (2002) 1423–1439.
[19] Y. Lu, Z. Tu, L.A. Archer, Nat. Mater. 13 (2014) 961–969.
[20] J. Xiu, L. Ma, J.R. Dahn, J. Power Sources 287 (2015) 377–385.
[21] M. Nagahama, N. Hasegawa, S. Okada, J. Electrochem. Soc. 157 (2010) A748–A752.
[22] H. Duncan, N. Salem, Y. Abu - Lebdeh, J. Electrochem. Soc. 160 (2013) A838–A848.
[23] Y. Yamada, et al., J. Phys. Chem. C 114 (2010) 11680–11685.
[24] R. Hayes, G.G. Carr, R. Atkin, Chem. Rev. 115 (2015) 6357.
[25] R. Youriesi, M. Hahlin, M. Robertson, K. Edstrom, J. Power Sources 225 (1) (2013) 40.
[26] M.Y. Nie, D. Chalsani, D.P. Abraham, Y.J. Chen, A. Bose, B.L. Lucht, J. Phys. Chem. C 117 (2013) 1257–1267.
[27] B. Li, Y. Wang, W. Tu, Z. Wang, M. Xu, L. Xing, W. Li, Electrochim. Acta 147 (2014) 636–642.
[28] G. Yang, et al., RSC Adv. 7 (2017) 26052.
[29] A. Wang, S. Kadam, H. Li, S. Shi, Y. Qi, Comput. Mater. 4 (2018) 15.
[30] H.J. Bergveld, W. Krujit, P.H. Notten, Springer (2002) 107–108, 113.
[31] S. Dhameja, Newnes Press, 2001, p. 12.
[32] H.C. Choi, Y.M. Jung, I. Noda, S.B. Kim, J. Phys. Chem. B 107 (24) (2003) 5806–5811.
[33] A. Sendek, et al., Energy Environ. Sci. 10 (2017) 306.
[34] W.J. Zhang, J. Power Sources 196 (2011) 13.
[35] H. Wu, Y. Cui, Nano Today 7 (2012) 414.
[36] H. Wu, et al., Nat. Commun. 4 (2013).

CHAPTER 2

Background and design of battery separators

Introduction

The structure and properties of the separator material strongly affect the overall battery performance. This is because, although the separator itself does not actively participate in the operation of the battery, the presence of the separator in a battery adversely affects the overall battery performance, because it adds electrical resistance and occupies space. The separator should be as thin as possible and porous, but it needs to possess the physical strength to maintain separation, both mechanical and electrical, between the electrodes. These features are essential to achieve the high energy and power densities for use that require such attributes. Therefore, the practical design of a lithium-ion battery separator is a compromise between the level of porosity of the separator and its mechanical strength. A comprehensive review of battery separators with a primary emphasis on lithium-ion—based technologies has appeared in the literature [1].

The ideal battery separator would be infinitesimally thin, offer no resistance to the transport of ions in electrolytes, provide infinite resistance to electronic conductivity for the isolation of electrodes, and be highly tortuous to prevent dendrite growth. In addition, it would be inert to chemical reactions. However, in the real world, the ideal case does not exist. Real separators are insulating membranes whose ionic resistivity is brought to the desired range by manipulating both the membrane thickness and its porosity. It is ultimately the application that decides which separator is most suitable. In that regard, it is clear that no single separator meets all needs of battery designers; some compromises need to be made.

This chapter summarizes some key properties of lithium-ion battery separators and how those properties can be manipulated to meet specific needs. The relation between the separator properties and battery performance will be discussed in detail. This is one aspect of separator properties that is not always highlighted in the technical literature.

Polymer-Based Separators for Lithium-Ion Batteries
ISBN 978-0-12-820120-6
https://doi.org/10.1016/B978-0-12-820120-6.00005-2

In addition to separator properties in isolation, the properties in an actual battery environment are important. In particular, the effect of the presence of the electrolyte on the properties of the separator needs to be further investigated. In addition, the forces that are present in the battery affect the properties of the separator, and these effects are discussed as well. Further characterization techniques that can help identify changes in the separator that occur during actual battery operation need to be developed. Finally, structure—property relations for the separator are defined. Those structure—property relationships can be used to develop the necessary separator structure for emerging applications. A major part of that work is to realize that the same separator will not necessarily be the optimum structure for all emerging applications. A single structure may need to be adjusted as needed, or new structures may need to be developed for these novel uses for lithium—ion batteries.

Separator properties

The separator is a critical component of a battery. It provides a barrier between the anode and the cathode while allowing the exchange of ionic charge carriers from one side of the battery to the other. The structure of a separator in a lithium-ion battery is shown in Fig. 2.1. As noted, the separator performs two important functions that are vital to the performance of the battery.

Many commercially available lithium-ion batteries use polyolefins as the primary separator material. A polyolefin is a class of polymer produced by the polymerization of an olefin monomer. In general, the olefin monomer comes from a petrochemical source. Polyolefins have excellent mechanical properties and good chemical stability, and they are relatively low in cost compared with other materials.

One important property of a battery separator is its porosity. The porosity is the ratio of the void volume to the apparent geometric volume. Typical commercial separators for use in lithium-ion cells have a porosity of about 40%. In addition, a uniform pore size distribution is desirable within the total area of the separator film to achieve a uniform current distribution. Furthermore, small, submicron pore-size dimensions have been shown to be effective in reducing the potential of an internal short circuit by preventing the penetration of larger active electrode material particles.

The challenge of designing efficient battery separators is the trade-off between mechanical robustness, as measured by the mechanical properties,

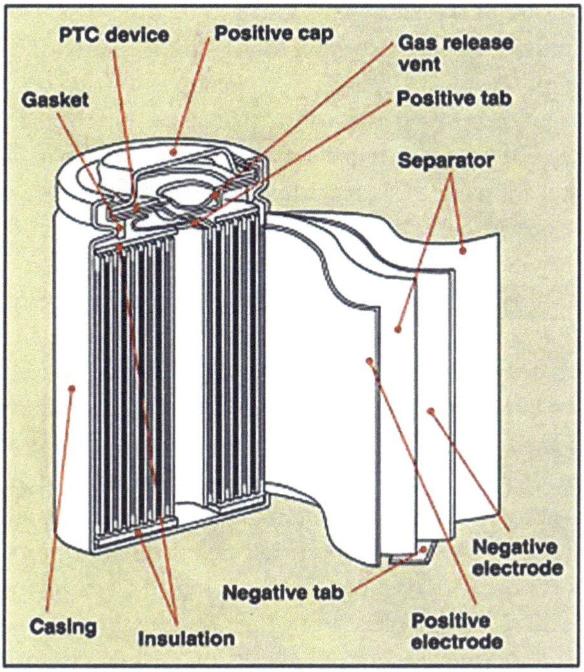

Figure 2.1 Structure of lithium-ion battery with separator. *PTC, positive temperature coefficient.*

and the porosity or transport properties that are required of the separator. As the porosity of the separator increases, the mechanical properties will typically decrease. Thus, the two features of the separator have opposite trends. The balance of these two features is critical to the optimum performance of the separator in a battery.

The overall mechanical strength of the separator is defined by several factors, each of which possesses desirable needs and performance criteria. The film tensile properties highly depend on the manufacturing process used for its production. High mechanical strength is important, especially in the separator film's machine direction, because the separator is subjected to considerable tension in this direction during the winding process of a spiral-wound battery and contraction of the width during that operation is highly undesirable.

Puncture strength is a measure of a material's resistance to penetration from a sharp object or through a blunt force. High puncture strength values are important to prevent the penetration of particulate electrode material through the separator. This is particularly important during the assembly of lithium-ion batteries.

The first feature of the battery separator to be discussed is its thickness. A minimal separator thickness is required for high-energy and high-power applications. However, this can have a negative impact on the separator mechanical strength and on the safety that it provides. For a long time, commercial separators for lithium-ion cells were less than 30 μm thick. Emerging applications such as electric vehicle, are considering the use of separators that are 10 μm and thinner.

Thin battery separators provide several advantages that can improve battery performance and design. First, thinner separators enable higher rate capacities to be attained. This means that the battery is able to produce higher current densities across the membrane because the electrical resistance of the separator can be lowered by reducing the separator's thickness. Higher rate capabilities are important where quick power is needed, such as in acceleration in an electric vehicle. Second, thinner separators allow for the use of thinner electrodes. This translates into the battery attaining better electrode use and cycling. Electrode use means that smaller batteries with equivalent power can be produced. Better electrode cycling means a greater number of recharges or cycles over the lifetime of the battery can be achieved. Third, thinner separators allow for greater choices in separator design. In other words, various thin separators can be combined to custom design a separator for a particular battery.

The pore size in battery separators must be smaller than the particle size of the electrode components, including both active materials and any conductive additives. Ideally, the pores should be uniformly distributed while providing a tortuous structure. This guarantees uniform current distribution through the separator while inhibiting the growth of lithium on the anode. The distribution and structure of pores can be analyzed using a capillary flow porometer [2]. A porometer measures the pore size of the separator's through pores, which have an important effect on its behavior and transport properties.

A critical component of the function of the separator is its permeability. Permeability is a measure of the ability of a porous material to transmit a fluid such as an electrolyte. The presence of a solid separator material increases the effective resistance of the ion-conducting electrolyte. The permeability of a separator is related to the material's porosity and tortuosity, which is the ratio of the mean effective capillary length to the separator thickness. Electrical resistance [3] can be used to gauge the performance of the separator in an actual electrochemical cell environment. The data obtained from such measurements are useful performance indicators because they define voltage loss within the cell during discharge and provide a measure of rate capabilities.

For longevity purposes, the battery separator needs to have a high degree of chemical stability to resist degradation and loss of mechanical strength resulting from reactivity or dissolution by the electrolyte material. In addition, the separator needs to have electrochemical stability to highly oxidative or reductive environments provided by the active electrode cell components. Degradation of the separator material properties can be observed as an indicator of the effect of the chemical environment of the separator.

For battery safety purposes, a state-of-the-art battery separator should have the ability to shut down the battery if overheating occurs. This overheating can be caused by overcharging or abuse of the battery. When a threshold temperature is reached, reactions termed self-heating can occur owing to exothermic reactions that occur internally within the battery [4]. Such reactions can include events that occur between lithium and the electrolyte and also the thermal decomposition of internal battery components. If internal heat generation is allowed to continue, a catastrophic event known as thermal runaway could occur. This is a serious safety concern.

The shutdown property of a polymeric separator provides an internal margin of safety against such events as an external short circuit, accidental overcharge, or abuse situation that results in an elevated cell temperature. Near the melting point temperature of the specific polymeric material providing shutdown, the pores collapse and close. This effect is accompanied by a dramatic increase in overall cell impedance, resulting in an end of electrochemical activity and current flow. Overall optimization of the shutdown effect of the separator can be achieved by tailoring the material properties and processing methods, and by using additives.

A battery separator should also display a high-temperature melt integrity, which involves showing mechanical robustness above the separator shutdown temperature. After a shutdown event occurs, the battery temperature is likely to continue to increase. The separator needs to maintain mechanical integrity and high impedance as a function of time at elevated temperatures. This is necessary to prevent the electrodes from making physical contact with each other and to create the safety issue of an internal short circuit.

Separator properties during battery performance

Not only the properties of the separators in the initial state but also the properties during the actual battery operation are important. The mechanical properties of polymeric-based separators depend on the temperature, loading rate, and loading history [5]. The failure of separators under abusive loading conditions will lead to internal short circuits and potential thermal runaway. However, the exact mechanism of short circuits and the conditions that lead to thermal runaway are still incompletely understood. Local penetration experiments on pouch cells have yielded puzzling results. Some batteries experience short circuits and thermal events under predictable levels of force and displacement whereas other cells withstand much larger deformations with no noticeable failure [6,7]. Two different failure mechanisms under biaxial loading for commercial membranes were identified [8]. That work showed that there is a significant difference in the size of the short-circuit area under the two different failure modes. An important implication of these findings is that it provides a possible explanation of the extent of electrochemical and thermal reactions that are observed when each mode is activated.

Peabody et al. [9] investigated the effect of viscoelastic creep of battery separators on the performance of lithium-ion batteries. They demonstrated that creep reduces ion transport via pore closure, resulting in increased internal resistance. This means there is an important link between the mechanical properties of the separators and the overall electrochemical performance of the battery. Elevated temperatures will further increase the creep rate as well as the mechanical effects.

Gor et al. [10] studied the effects of electrolyte solvents on the mechanical properties of a battery separator through both thickness and elastic modulus measurements of samples immersed in different solvent environments. Certain electrolyte solvents such as dimethyl carbonate, diethyl carbonate, and ethyl acetate cause significant softening of the separator. On the other hand, in other solvent environments such as propylene carbonate and acetonitrile, the separator material retained the properties of the dry material. The reduction in the mechanical property is attributed to polymer swelling; the results are explained in the context of solubility and interaction parameters.

Compressive deformation has been shown to have an impact on the ionic transport characteristics of separators [11]. The deformed pore networks lead to a decrease in ion transport and an increase in battery impedance. The increase in impedance associated with the separator can be explained by porosity and thickness changes that occur during deformation. A relationship known as the Bruggeman relation (Eq. 2.1) is a good description of the tortuosity of a separator undergoing compression deformation:

$$\tau = \zeta \varphi^{1-a} \qquad (2.1)$$

In this equation, τ is the separator tortuosity, ζ is a constant, φ is the separator porosity, and a is the Bruggeman exponent [12]. Cannarella and Arnold [11] provided an experimental measurement of Bruggeman parameters for various separators undergoing compression.

It was found [13] that the behavior of a polypropylene separator in compression is different from that under tension. This shows that the tensile properties cannot necessarily be used as a substitute for the compressive properties of a separator. In particular, the softening effects resulting from polymer–solvent interactions for separators are much more pronounced under tension than under compression. This further implies that knowledge of the compressive properties of separators is necessary to understand the longtime performance of separators in real battery cells where compressive loads are operating.

These studies investigated the in situ effects of compressive deformation on the performance of polymer battery separators. Earlier research demonstrated through ex situ electrochemical studies on transversely compressed pouch cells that mechanical stress can result in separator deformation and a significant degradation of cell performance [9]. The importance of the effects of mechanical and thermal stimuli on separators has been the motivation for much work on the mechanical behavior of separators [14,15]. Such studies have been become increasingly important because battery cells are desirable for long service life and operation at higher temperatures. The use conditions provide environments in which substantial separator deformation can occur as a result of mechanical creep. A primary message of the studies that has been quoted the features of the battery separator as initially made are important and significant, but it is also relevant to know how those properties change as the lithium-ion battery operates. Of course, measurement of these changes can be difficult because separator changes occur in situ during actual battery function. An approach to use is to disassemble the battery after a certain period of operation. However, such studies suffer from the fact that testing can be done only at certain intervals, and it is possible that events that cause changes in the separator may have occurred in a different time frame. Also, disassembling the battery must be done carefully to ensure that the actual process of taking the battery apart does not cause changes to be initiated in the various battery components, including the separator.

These comments point to the need to develop additional characterization techniques that monitor separator behavior under the conditions of a typical battery operation. The best approach in this regard would be to have testing procedures that can follow the separator in situ when the battery is working. However, such tests will be difficult to develop. The next best approach is to subject the battery separator to typical conditions that it experiences in a battery. Those conditions need to involve both thermal and stress situations that are encountered. The performance of the separator needs to be monitored and changes should be noted. It is likely that changes in overall battery performance can be linked to events occurring in the separator.

One aspect of a battery separator that has not yet been discussed is the degradation of the separator during charge and discharge cycles of the battery during its operation. Zhang et al. [16] studied dry process trilayer separators from disassembled fresh cells and cells with various charge–discharge cycles. They found that the separator changed color with higher cycles and was almost brown after 1200 cycles. These changes lead the mechanical strength of the separator to experience degradation as the battery cycles.

This effect for the separator translates to lower strength at the battery level. Thus, the structural designs of a battery should consider safety problems caused by the effects of aging that occur during charge and discharge cycles. Among the factors that can influence the aging of battery separators are chemical oxidation of the film itself, an increase in resistance caused by the accumulation of chemical particles in the pores, and the effect of the electrolyte.

In that regard, a review of the effects of the separator structure and chemistry on lithium-ion battery performance was provided [17]. One conclusion of that review is that the overall performance of lithium-ion batteries can be improved by considering the interplay of the separator with the surroundings, including the electrolyte. This view is different view from assuming that the separator acts in isolation and is not an active component in a battery. Instead, the separator influences the rate performance, cell life, and safety. As such, in the future, separators will be designed to have a more active role in the lithium-ion battery operation.

In general, the overall stability of lithium-ion battery separators under potentially extreme battery conditions will require leading-edge design and fabrication techniques to exceed manufacturing and end-performance requirements for large-scale applications. Those requirements include increased reliability and safety as the size of the lithium-ion batteries continues to grow to meet demands. Separator designs to guarantee safety should be considered for a specific battery size, application, and abuse situation. Separators that are acceptable for mobile phone batteries may or may not be the correct choice to guarantee safety in a plug-in hybrid or utility storage battery. New separator designs, detailed modeling, and robust testing are all essential elements to ensure that the selected separators meet both the performance and safety requirements for a given application.

One way in which the thermal stability of lithium-ion separators has improved is in the coating of the base substrate with a ceramic layer [18]. The addition of the ceramic layer prevents short circuiting within the lithium-ion batteries while maintaining mechanical integrity at elevated temperatures. An extensive discussion of the details of the ceramic coating process will be provided in a later chapter in this book. However, the development of ceramic-coated separators is a partial realization that separators that were initially developed for consumer electronics applications may not be the most effective separator material for use in emerging applications such as electric vehicles. This is an example of the need to understand the overall performance of lithium-ion battery separators in new applications.

Consumer electronics versus electric vehicles

As mentioned, separators that operate effectively in consumer electronics applications may not be the best materials for other uses, such as electric vehicles. A primary reason is the larger size of the batteries in these other applications. The larger batteries lead to the increased potential for thermal runaway events to occur. A review of the thermal runaway mechanism for lithium-ion batteries in electric vehicles is available [19].

Within the automobile industry, manufacturers use different types of batteries. For example, Tesla is committed to automobile batteries made from multiple cylindrical cells that look like slightly enlarged AA batteries, On the other hand, the Chevrolet Volt uses larger battery pouch cells in their electric vehicles. The different shapes and sizes of the batteries can require separators of different compositions. It is not clear what type of separator will dominate the electric vehicle market.

A potential drawback of lithium-ion batteries in electric vehicles is sensitivity to temperature. In addition, low-temperature power performance and performance degradation with time are downsides. Owing to the volatility of organic electrolytes, increases in temperature can lead to the combustion of these materials, which can cause fires and ultimately explosions.

Electric vehicle batteries are different from consumer electronics batteries. They are designed to be more robust, because they carry a heavier load and last much longer. Of course, they are also a little more expensive than the average battery for use in consumer electronics applications. In addition, electric vehicle batteries also tend to be large and inefficient compared with consumer electronics batteries. This is why today's lithium-ion batteries, which are found in nearly all consumer electronics, will require additional technological advances before they can be applied widely to tomorrow's electric vehicles. Among necessary improvements are those of safety, durability, and cost reductions.

Advances in nanotechnology for lithium-ion batteries will likely contribute to other areas of innovation including fuel cells, electronics, and biotechnology. The United States is a leader in nanotechnology development. Companies are using nanotechnology expertise to enter the market for lithium-ion battery materials. These developments will be necessary to allow lithium-ion batteries to be used effectively in electric vehicles. Nanomaterial technology needs to move to the high-volume production stage, which will produce benefits not only for lithium-ion battery technology but also for many other industries.

Nanotechnology can have a significant role in achieving specific performance objectives in electric vehicle batteries. Conventionally, graphite powder has been used in the negative electrode of lithium-ion batteries. It serves as an intercalation material in this application. However, the rate of removal or insertion of lithium and the battery capacity can be improved by replacing the micron-sized powder with other carbon materials such as carbon nanotubes [20]. Because carbon nanotubes have a high surface area, they can bind much higher concentrations of lithium. Nanowires made of titanium dioxide or tin oxide are also promising candidates as negative electrode materials.

In terms of separator materials, the structure that is optimal for use with graphite powder may not be best with these new nanomaterials. The size of the particles in the nanomaterials can raise concerns about the potential for punctures occurring in the separators. This concern becomes a larger issue for thinner separators because they possess lower puncture strength values and thus are more susceptible to the production of holes.

The use of nanomaterials also has the effect of improving the current density. This is because the diffusion path for lithium ions is reduced, which increases their mobility. In addition, the use of nanomaterials allows for a higher electrode–electrolyte contact area from the inherently high surface area features of the particles [21]. Also, reducing the particle size in the electrodes is believed to reduce mechanical stresses, which are caused by volumetric expansion and contraction during charge and discharge processes.

These advances can lead to significant developments in the battery separator used in these batteries. For example, features of the separator will need to be modified to allow for the rapid transport of lithium ions. Those modifications should enhance lithium-ion conductivity improvements that can be observed. In addition, the changes may lead to improved rate capability and cycling durability in lithium-ion batteries.

This work highlights the fact that the overall function of the separator in electric vehicle applications will likely be different from what it is in traditional consumer electronics uses. Traditionally, the separator prevents electronic conduction and allows for easy ionic transport. However, functional separators may need to have additional features such as high thermal stability, high rate operation, lithium dendrite suppression, and flame retardation. The separator can possess an ordered porosity and may consist of a hybrid structure with functional inorganic nanomaterials. In addition, it may contain functional polymers or a polarity-controlled surface. The deposition of functional materials on conventional separators is a facile method to achieve that goal. Using that approach, tailored separators can be manufactured that meet the needs presented by the electromobility market.

An approach that uses nanoparticles in battery separators that needs further exploration is discussed in Kim et al. [22]. In that work, inorganic nanoparticles with a hierarchical pore structure were coated on a conventional polymer separator. The resulting separator showed superior lithium-ion transportation compared with separators coated with mesopore-only or conventional nonporous nanoparticles. The presence of the mesopores and micropores acts synergistically to improve the uptake of the electrolyte and the ionic conductivity of the inorganic particles. This is achieved while positive features, such as their thermal and mechanical properties, are maintained.

Other emerging applications

Along with hybrid electric vehicles, plug-in hybrid electric vehicles, and pure electric vehicles, a growing application for lithium-ion batteries is their use in smart grid applications. In general, the overall battery size for use in these applications is similar to that of batteries used in electric vehicles. Therefore, many of the same issues are present in the two types of batteries. For example, the safety of the batteries is a critical aspect of their behavior. The separator can be a critical part of this feature by keeping the electrodes apart to prevent short circuits. In addition, the lifetime of the battery is an important factor in power grid applications. This is because batteries used in such applications must function effectively for a long time. The separator can have an impact on that feature of the battery by preventing the pores from clogging enabling the continued and efficient transport of lithium ions. Some features that nanomaterials can provide were discussed earlier in this chapter for electric vehicles. Because lithium-ion batteries for grid applications are similar in many ways to those used in electric vehicles, identical arguments can be made here. In particular, the development of separator materials for these emerging power applications will likely need to follow a path similar to that for developments occurring in electric vehicles. Any safety enhancement that the use of nanomaterials can provide to batteries and separator materials is expected to be quickly implemented.

There are some differences between batteries for electric vehicles and grid storage use. For example, batteries for transportation applications need to be relatively compact and they require high volumetric energy and power densities. These factors are less critical for power grid storage, because the footprint is not often a limiting criterion. If a site has needs additional storage capacity, it can often be met by adding an container of batteries. However, such a situation would make battery safety an even more important feature. The separator is a critical component of the battery to assist in achieving this goal.

Thus, the lithium-ion battery needs to show constant improvement to meet the needs of these emerging applications. Scientists are experimenting with new ways to combine electrolytes, anodes, and cathodes to create a battery that is more energy-efficient, more cost-efficient, and safer than its current form. The separator and its contribution to these effects are often overlooked, and it is not clear whether the separator used in current batteries is optimal for use in these developing battery technologies. Future separator developments need to consider this and address the needs of the new markets appropriately.

Conclusions

Factors affecting the overall design of battery separators used in lithium-ion batteries were discussed in detail. Relationships between the design of the separator and its performance in the battery environment were highlighted. The correlation of separator properties and the overall performance of the battery was a key topic of this chapter. This is an aspect of separator properties that has not always been highlighted in the literature on this topic.

In that sense, a major conclusions of the chapter is that separator properties in isolation are important, but the properties in the actual battery operating environment also need to be further considered. Effects of the presence of the electrolyte on battery separator performance need to be considered. Also, the influence of forces present in the battery need to be considered when discussing separator properties. Additional characterization techniques need to be developed to measure separator properties effectively during actual battery operation. Finally, the effects of electrochemical reactions that occur during battery operation need to be taken into consideration when designing battery separators.

Overall, there needs to be a complete understanding of the structure—property relationships of separators and how those relationships affect lithium-ion battery performance. There is more to the separator situation than simply stating that it is based on either polypropylene or polyethylene. Rather, the structure of the separator itself in terms of porosity, pore size, and pore size distribution, as well as other properties, needs to be correlated with battery performance. The optimal range of parameters for each of these properties needs to be defined to use the separator in a battery for a specific application.

Furthermore, the best separator for use in consumer electronics applications may not be best in emerging applications such as electric vehicles and power grid applications. The needs of each application need to be examined in detail to determine the separator structure that best satisfies the requirements. Also, as new electrode and electrolyte materials are developed, the need to optimize the separator structure to use those novel formulations best should be considered.

Certain separator features make them attractive for use in certain applications. On the other hand, some characteristics indicate that the separators may not be best for use in other applications. A complete understanding of these features is essential to the further development of battery separator materials for new and emerging applications such as electric vehicles and energy storage. Without that understanding, there is no guarantee that the separators and the batteries that contain them will operate at the most effective level. This could have a significant impact on the optimal implementation of the new materials and associated technologies.

References

[1] S. Zhang, J. Power Sources 164 (2007) 351.
[2] [2a] K.M. Abraham, Electrochim. Acta 38 (1993) 1233. [2b] Capillary Flow Porometer Reference.
[3] P. Arora, Z. Zhang, Chem. Rev. 104 (2004) 4419.
[4] J. Dahn, et al., Solid State Ionics 69 (1994) 265.
[5] X. Zhang, Mechanical Behavior of Shell Casing and Separator of Lithium - Ion Battery, Massachusetts Institute of Technology, 2017.
[6] E. Sahraei, J. Campbell, T. Wiezbicki, J. Power Sources 220 (2012) 360.
[7] X. Feng, et al., J. Power Sources 275 (2015) 261.
[8] X. Zhang, Sci. Rep. 6 (2016) 32578.
[9] C. Peabody, C.B. Arnold, J. Power Sources 196 (2011) 8147.
[10] Gor, et al., J. Power Sources 294 (2015) 167.
[11] J. Cannarella, C.B. Arnold, J. Power Sources 226 (2013) 149.
[12] I.V. Thorat, D.E. Stephenson, N.A. Zacharias, K. Zaghib, J.H. Harb, D.R. Wheeler, J. Power Sources 188 (2009) 592.
[13] J. Cannarella, X. Liu, C.Z. Leng, P.D. Sinka, Y. Gor, C.B. Arnold, J. Electrochem. Soc. 161 (11) (2014) F3117–F3122.
[14] A. Sheidaei, X. Xiao, X. Huang, J. Hill, J. Power Sources 196 (2011) 8728.
[15] C.T. Love, J. Power Sources 196 (2011) 2905.
[16] X. Zhang, J. Zhu, E. Saharaei, RSC Adv. 7 (2017) 56099.
[17] M.F. Lagadec, R. Zahn, V. Wood, Nat. Energy 4 (16) (2019) 18. U.S. Patent 6,432,586, Assigned to Celgard, LLC.
[18] C.T. Love, J. Power Sources 197 (2013) 3143.
[19] X. Feng, et al., Appl. Energy 246 (2019) 53.
[20] P. Sehrawat, C. Julian, S.S. Islam, Mater. Sci. Eng. B 213 (2016) 12–40.
[21] A. Manthiran, A.V. Murugan, A. Sarkar, T. Muraliganth, Energy Environ. Sci. 1 (2008) 62.
[22] Y. Kim, T.P. Tran, M. Kim, D.W. Jung, G. Yi, J.H. Park, ACS Appl. Mater. Interfaces 7 (8) (2015) 4511–4517.

CHAPTER 3

Characterization techniques for battery separators

Introduction

The separator is a permeable membrane placed between a battery's anode and cathode. The main function of the separator is to keep the two electrodes apart to prevent electrical short circuits while allowing the transport of ionic charge carriers needed to close the circuit during the passage of current in an electrochemical cell. As such, the separator must have certain properties that allow it to function effectively. There are methodologies and procedures typically used to measure and characterize these properties. This chapter will discuss those test procedures in detail and provide some examples of typical data obtained using the indicated tests.

Separators can be characterized with two different types of properties. The first are structural properties that describe what the separators are, and the second are functional properties that discuss how the separator performs. Examples of structural properties include the chemical nature, thickness, pore size, pore size distribution, porosity, and properties such as chemical stability and electrolyte uptake. Functional properties include the electrical resistance (ER), permeability, and transport number. A variety of techniques are used to evaluate battery separators and measure these properties. Many of the techniques will be discussed in this chapter.

In addition to the properties of separators in isolation, it is important for characterization techniques to be developed and implemented that define separator features during actual battery operation. The effect of the presence of the electrolyte and the electrode materials on separator characteristics needs to be better defined. Also, the impact of the compressive forces present in the actual battery environment needs to be considered.

Polymer-Based Separators for Lithium-Ion Batteries
ISBN 978-0-12-820120-6
https://doi.org/10.1016/B978-0-12-820120-6.00004-0

Structural properties

Thickness

The thickness of battery separators is usually measured by ASTM D5947-96 [1]. Lithium-ion cells employed in consumer applications generally use separators less than 25 μm thick. In general, some separators being developed for electrovoltaic applications are thicker and are about 40 μm thick. The thicker the separator is, the greater is the mechanical strength and the smaller is the potential of punctures during cell assembly. On the other hand, thicker separators permit the inclusion of less active material in the battery. The thinner separators occupy less space and allow for the use of longer electrodes. These effects lead to an increased capacity as well as rate capability. In addition, thinner separators add less resistance to the battery.

An important parameter related to the average separator thickness is the spread in the thickness values that are measured. This is significant because the thickest sections of the separator will ultimately determine the amount of space required in the battery for the separator. To obtain a reasonable estimate of variations in thickness in the separator, multiple thickness measurements should be made in various locations of the separator. A desirable standard deviation in thickness measurements is less than 10% of the total thickness.

Porosity

The porosity of the separator is an important property for high permeability and for the provision of electrolyte in the battery. High and uniform porosity is a desirable feature for the unhindered flow of ionic current. On the other hand, nonuniform porosity leads to nonuniform current density. This can lead to reduced activity of the electrodes. The result can be cell failure because the discharge of some areas of the electrodes works harder than other areas during battery operation.

The porosity of a separator is generally defined as the ratio of the volume of voids to the apparent geometric volume. It can be calculated from Eq. (3.1):

$$\%\ \text{porosity} = [1 - (\text{sample mass}/\text{sample volume})/\text{polymer density}] \times 100 \quad (3.1)$$

Usual input parameters into Eq. (3.1) are the polymer density, basis weight, and dimensions of the material. Using this approach to calculate the percent porosity does not distinguish between accessible and inaccessible portions of the separator material.

One of the standard test materials for determining the porosity is ASTM D-2873 [2]. If there is interest in determining the accessible porosity, the weight of liquid that is absorbed in the pores of the separator can be measured. A typical liquid used in these experiments is hexadecane. In this test, the weight of the separator is measured before and after dipping in hexadecane. The porosity is thus calculated according to: Eq. (3.2):

$$\%\text{porosity} = \frac{\text{volume occupied by hexadecane}}{\text{volume of polymer} + \text{volume occupied by hexadecane}} \times 100 \tag{3.2}$$

This approach assumes that the volume that is occupied by the hexadecane is equal to the pore volume of the separator.

Pore size and pore size distribution

For use in batteries, the separator should have a uniform pore size distribution. This is desirable to avoid nonuniform current densities that can lead to performance loss in the battery. Pores in the submicrometer range are crucial for presenting internal short circuits between the anode and the cathode of the lithium-ion battery. This is particularly important because most separators are as thin as 25 μm or less. This aspect of the separator becomes more increasingly important as separators become thinner to increase cell capacity.

Testing of battery separators and control of their pore features are important characteristics for the proper functioning of lithium-ion batteries. Historically, mercury porosimetry was used to characterize separators in terms of percent porosity, average pore size, and pore size distribution [3]. In this approach, the size and volume of pores in a separator are measured by determining the amount of mercury that can be forced into the pores while increasing the pressure. However, mercury does not wet all materials and an external force must be applied to the separator to overcome the surface tension forces that oppose entry of the mercury into the pores.

Another technique that can be used with separators is the Aquapore technique, in which water is used in place of mercury. Porosimetry provides pore volume, surface area, mean pore diameter, and pore size distribution [4]. As such, useful information can be obtained to characterize polyolefin-based separators commonly used in lithium-ion batteries.

In a typical Aquapore experiment, the sample is placed in the equipment and evacuated. The pressure is then slowly increased. As the pressure increases, the amount of water that is forced into the pores increases in direct proportion to the differential pore volume. The differential pore volume is related to the size of the pores that correspond to the instantaneous pressure. Therefore, increasing the pressure for a separator with a certain pore size distribution provides a unique volume versus pressure or pore diameter curve. The pressure needed for the intrusion of water into a pore of diameter D is provided by the equation:

$$D = 4\gamma \cos\theta / P \tag{3.3}$$

In this equation, D is the diameter of the pores if they are assumed to be cylindrical, P is the differential pressure, γ is the surface tension of water, and θ is the contact angle of water. The pores usually vary in form and size and are generally not spherical or a constant diameter.

Another method, called capillary flow porometry, was developed by Porous Materials, Inc. [5] to characterize the pore structure in battery separators. This technique [6] uses samples with pores filled with a wetting liquid. The gas pressure required to remove the liquid from the pores and the flow rate of the gas through open pores are measured. Excellent reproducibility in determining the intrusion volume, porosity, and pore diameter can be obtained with this technique. In addition to those parameters, the size of the pore at its most constricted part and the largest pore size can be determined.

Scanning electron microscopy can also be used to investigate the pore structure of battery separators [7]. In that work, because the separators are not conductive, they were coated with gold particles for 1 min in a sputter coater before the images were observed. The study showed that the pores of commercial separators are uniformly distributed.

Shrinkage

Shrinkage testing is normally carried out in both the machine direction (MD) and transverse direction (TD). In this test, the separator dimensions are measured and the separator is then stored at a particular temperature for a fixed time. Typical temperatures for performing shrinkage testing are 90 and 150°C. After the separator is exposed to elevated temperatures, the dimensions of the sample are measured again. The shrinkage is calculated from the change in dimensions, as shown in Eq. (3.4):

$$\text{Shrinkage}(\%) = (L_i - L_f)/L_i \times 100 \tag{3.4}$$

where L_i is the initial dimension and L_f is the final dimension of the separator after exposure to elevated temperatures. Uniaxially stretched separators tend to shrink only in the MD, whereas biaxially stretched separators shrink in both the MD and TD. The shrinkage of separators can also be measured by thermomechanical analysis (TMA) and dynamic mechanical analysis (DMA) under conditions of constant load and rate [8].

Puncture strength

It is necessary for a separator to have adequate physical strength to withstand the cell assembly process and daily charge–discharge cycling. Physical strength is required to withstand basic handling procedures, cell assembly, physical shock, punctures, abrasion, and compression. These are all examples of forces and situations the separator may encounter during battery fabrication and operation processes.

The puncture strength of a separator is a measure of the weight that must be applied to a needle to force it completely through a separator [9]. Puncture strength has been used to show the tendency of separators to permit short circuits in a battery. These short circuits may occur as the result of holes generated in the separator by the uneven surface of an electrode during battery assembly and charge–discharge processes. Reproducible results for puncture strength values can be obtained by using a load frame such as an Instron machine.

The puncture strength of the separator depends on the materials used to produce the separators as well as the manufacturing method. Separators that are biaxially oriented have high tensile modulus and rupture strength values in both directions. Features such as high polymer entanglements and orientation can increase the puncture strength of the separator.

Tensile strength

Tensile strength testing is done in both the MD and TD. As with puncture strength, tensile properties depend on the manufacturing process. Uniaxially oriented membranes have high strength values in only one direction, whereas biaxially oriented membranes have strength in both the MD and the TD. ASTM D88-00 [10] is the appropriate test to determine tensile strength.

The separator should be strong enough to withstand mechanical handling during cell assembly. It should possess dimensional stability and not experience neckdown during winding because such neckdown will cause for a decrease in the film width. If this occurs, the decrease in width will permit electrodes to contact each other and the result may be a short circuit to the battery.

Electrolyte uptake

Any good separator should be able to absorb a significant amount of electrolyte and retain the absorbed electrolyte when the battery is in operation. These effects are more important in sealed cells in which no free electrolyte is present. A maximum amount of electrolyte in the separator is highly desirable to achieve minimal cell internal resistance.

Separator wettability can affect the overall performance of batteries by increasing separator and cell resistance values. Separator wetting can be correlated with the electrolyte filling time in lithium-ion batteries. The wetting speed is related to the type of polymer used in the separator and the pore size and porosity. There is no accepted test procedure to determine separator wettability. However, placing a drop of the electrolyte material on the separator and observing whether the drop quickly wicks into the separator is a good qualitative indicator of wettability. The contact angle is also a good measure of wettability.

The wettability of separators can be improved by incorporating chemical functional groups into a coated layer. For example [11], the introduction of benzimidazole, amino, and carboxyl groups to a terephthaldehyde cross-linked high-performance separator improved both the wettability and electrolyte uptake of the separator. Specifically, absorption of the liquid electrolyte increased from 90% for the control polyethylene (PE) separator to 181% for the modified separator.

In addition, wetting additives can be used to enhance separator wettability toward electrolytes. As an example [12], block copolymers of polyoxyethylene and polyoxypropylene are reported as wetting additives. In addition, that work cited the use of water as a wetting additive to improve separator wettability toward traditional carbonate electrolyte materials.

Functional properties

Gurley

Separator permeability is usually characterized by air permeability. The Gurley number is a common parameter used to define air permeability. It expresses the time needed for a specific volume of air to pass through a specific area of the separator under a specific applied pressure. The standard method for measuring Gurley values is described in ASTM-D726B [13].

The Gurley value is often used to characterize separators, because the test is accurate and easily performed. In addition, it can be done in a fast time frame. Also, deviations from specific values are a good indicator of problems; thus, Gurley values can be used in quality control situations. Furthermore, it has been shown that the Gurley numbers are proportional to the ER for a particular separator morphology [14]. Because of that relation, Gurley values can be used in place of ER measurements once the mathematical behavior between Gurley and ER is established. A lower Gurley value means a higher porosity and accordingly a lower ER value.

Tortuosity

The tortuosity of a separator is the ratio of the mean effective capillary length to the separator thickness. The tortuosity factor, τ, is defined by the equation:

$$\tau = l_s/d \tag{3.5}$$

In this equation, l_s is the ion path through the separator and d is the thickness of the separating layer.

Tortuosity is a property of porous media. It is used to define τ by using electrical conductivity measurements. It is necessary to know the sample membrane resistance, thickness, area, and porosity to calculate the tortuosity factor.

The tortuosity factor value is often used to describe ionic transport by providing information about the effect of the pore structure. A tortuosity factor equal to 1 describes an ideal porous body, whereas values of $\tau > 1$ refer to systems that are more hindered. In general, higher tortuosity values indicate favorable resistance to dendrites, but they can also lead to higher separator resistance.

Electrical resistance

The determination of separator resistance is important because of the impact the separator has on electrical performance. Generally, ER is a more complete measure of permeability than is the Gurley value. This is because its measurement is carried out in the actual electrode solution. Because of this, it represents how the separator will function in the actual battery environment. It measures voltage loss within the battery and estimates rate limitations.

The ionic resistivity of the porous separator us essentially a measure of the resistivity of the electrolyte that is absorbed into the pores of the separator. Usually, a microporous separator that is immersed in an electrolyte has an ER value about six to seven times that of a comparable volume of the electrolyte. It is a function of the resistivity of the electrolyte, the thickness of the membrane, the membrane's porosity and tortuosity, and the extent to which the electrolyte wets the pores of the separators [15].

Classical techniques for determining the ER of battery separators were discussed in Falk and Salkind [16] and Robinson and Walker [17]. The resistivity of an electrolyte material is more accurately measured using AC methods because DC methods can cause electrolysis of the material. In addition, AC impedance measurements allow for quick measurement of the cell resistance over a wide range of frequencies. AC measurements yield important data about the long-range migration of ions and any polarization phenomena that occur within the battery.

In a typical AC measurement, a sinusoidal voltage is applied to the battery and the current that passes through as a result of this sinusoidal voltage is measured. For resistivity measurements, a four-electrode cell is usually employed. This approach is used because it avoids possible complications that can arise from a nonuniform potential field near the outer two electrodes.

The experimental procedure used to determine the ER involves cutting small pieces of the separator and placing them between two blocking electrodes. The separators are completely saturated with the appropriate electrolyte material. The resistance of the separator is then measured at a particular frequency using AC impedance methods. The actual frequency of the measurement needs to be chosen so that the separator impedance is equal to the separator resistance. To improve accuracy in the measurements, it is preferable to make multiple measurements by adding extra layers. In that approach, the average resistance of a single layer of the separator is determined using multiple measurements. The specific resistivity, ρ_s, is given by:

$$\rho_s = R_s A / l \tag{3.6}$$

where R_s is the separator resistance (in ohms), A is the area of the electrode (in cm^2), and l is the membrane thickness (in cm). In addition, the specific resistivity of the electrolyte, ρ_e, is given by:

$$\rho_e = R_e A / l \tag{3.7}$$

where R_e is the measured electrolyte resistance (in ohms) and A and l are the same parameters as before.

The ratio of the separator membrane resistivity to that of electrolyte resistivity is called the MacMullin number and is designated N_m. The MacMullin number has been used to predict the effect of the separator on overall battery performance. This is usually done through (Eq. 3.8) [18]:

$$N_m = \rho_s/\rho_e = \tau^2/\varepsilon \tag{3.8}$$

where τ is the tortuosity of the separator that was previously discussed and ε is the separator porosity. In general, the MacMullin number describes the contribution of a separator to overall cell resistance. Its value is essentially independent of which electrolyte is used and eliminates the thickness of the separator from the analysis. Its determination assumes that the electrolyte that is used in the test completely wets the separator.

Cannarella and Arnold [19] examined the increase in resistance associated with the deformation of the separator. The change in resistance depends on four parameters: strain, the Bruggeman exponent in the tortuosity relation [20], the initial separator resistance, and the initial separator porosity. The study showed that cell impedance increases with compression owing to decreases in ion transport. The increases in impedance associated with the separator can be attributed to changes in both porosity and thickness. Specifically, one of the most important factors associated with the induced increases in resistance is pore closure.

Mixed penetration strength

Mixed penetration strength is defined as the force required to create a short circuit through a separator owing to the penetration of electrode material. In the test itself, using a specific diameter ball, force is applied to the positive electrode—separator—negative electrode sandwich. The force at which the separator is penetrated and a short circuit is created is termed the mixed penetration force. Mixed penetration strength is used to show the tendency of separators to allow short circuits during battery assembly. The mix penetration test has been shown to be more closely related to particle penetration resistance than is puncture resistance [21].

Shutdown

Under conditions of rising temperature, pores of the separator become closed because of the melting process and the battery shuts down. For example, a PE separator shut the battery down when the core temperature reaches 130°C. The shutdown process stops ionic transport between the two electrodes. If the battery does not shut down as the temperature rises, heat in the failing cell could rise, which can lead to a thermal runaway effect, potentially causing the battery to catch fire.

When shutdown occurs, the porous polymer film is transformed into a nonporous insulating layer between the two electrodes. At the shutdown temperature, there is a significant increase in cell impedance and the passage of current through the cell is restricted. This effect prevents further electrochemical activity in the battery. The battery is shut down before a fire or explosion can occur.

The overall ability of the PE-based separator to shut the battery down is determined by its type, molecular weight, percent crystallinity, and process history. Thus, for example, high-density PE will show different shutdown features compared with low-density PE. Often, the material properties and processing methods need to be tailored so that shutdown is fast and complete. However, a challenge of this approach is to optimize shutdown features without a corresponding negative impact on the mechanical properties of the separator. A shutdown temperature of 130°C is often good enough to control cell heating and avoid thermal runaway in lithium-ion batteries. A lower shutdown temperature is desirable as long as it does not negatively affect the mechanical properties of the separator or the high-temperature performance. In fact, work has been performed to lower the shutdown temperature to 100–110°C [22]. The low-temperature shutdown was achieved in that work by applying ultrathin microporous coatings to the separators. A low-cost microporous coating less than 2 μm thick was proposed. The safety features developed should be particularly useful for large high-power batteries for which there is an increased potential for thermal runaway effects.

The shutdown feature of separators is typically measured by determining the separator impedance while the temperature linearly increases [23,24]. Fig. 3.1 shows a typical shutdown curve for a commercial separator from Celgard. The heating rate used in this experiment was about 60°C/min and separator impedance was measured at a frequency of 1 kHz. The rise in impedance that occurs starting at about 130°C corresponds to a collapse in the pore structure owing to the melting of the PE portion of the separator. A 1000-fold increase in impedance is required for the separator to stop thermal runaway in the battery. The drop in impedance observed at about 170°C results from opening of the pore separator. This effect can be caused by either the flow of polymer out of the pores or the penetration of the separator by the electrodes. This effect is referred to as a loss of melt

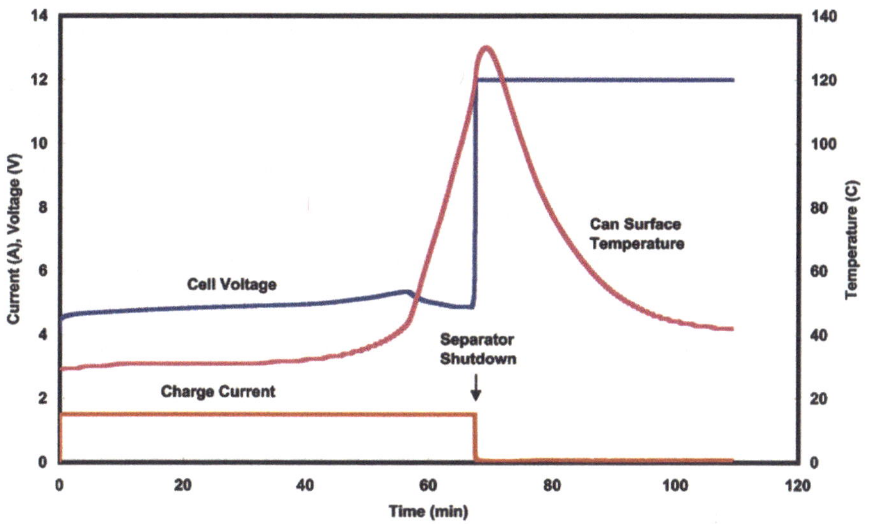

Figure 3.1 Shutdown curve for battery separator.

integrity. This property will be further discussed later in this chapter. The test is reliable for showing the temperature at which impedance begins to rise, but there is variation in the temperature at which a drop in impedance occurs.

Fig. 3.1 shows that the impedance rise occurred near the melting point of the PE at about 130°C. The shutdown temperature of the separator is governed by the melting point of the separator material. Thus, the shutdown temperature can be controlled and tailored by using materials with different melting points. This is primarily because pores in the separator collapse at the melting point, which leads to the formation of a nonporous film between the anode and cathode. The shutdown behavior observed in separators that are less than 20 μm thick is similar to that observed in thicker separators. Battery producers have been able to use thinner separators without much compromise in the shutdown behavior of the separators.

In situ small-angle X-ray scattering tests [25] have been used to study the shutdown behavior of separators. Shutdown of the pores was found to originate from the melting of crystals, leading to the release of the internal shrinkage force. Two driving forces were revealed for the shutdown effect: the shrinkage force from oriented polymer chains and the surface tension of nanosized pores. The work showed that the shrinkage force is the primary driving force at the initial stage and the surface tension controls the shutdown behavior at the final stage.

Laman et al. [23] introduced the experimental use of impedance as a function of temperature to characterize the shutdown behavior of separators. Employing a heating rate of 1°C/min, the researchers found that impedance increased several orders of magnitude near the melting portion of the separator. Spotnitz et al. [26] developed a thin layer cell that permits heating rates of 5°C/min and greater to be used.

As a way to tailor the shutdown temperature of a separator, work was done applying waxes on membranes [27]. In this work, a wax or low-melting polymer was coated on the polyolefin separator. The disadvantage of this approach is that the coating can block the pores of the separator. This can affect battery performance by increasing separator resistance. In addition, the coating layer needs to be high to achieve complete shutdown of the battery.

The shutdown feature provides protection from external short circuits and during overcharging of the battery. However, it provides a small amount of protection from internal short circuits. If the electrodes come in contact with each other or become shorted from the growth of a dendrite-soluble impurity or other dendrite material, the separator only assists in avoiding delayed failures. If an instant failure occurs during an internal short circuit, the rate of temperature increase is too high and the rate of the separator shutdown is not fast enough to control the thermal runaway effect from occurring.

Most shutdown separators were designed for the consumer market, which are generally low-voltage, smaller battery systems. If applied to higher-voltage, larger battery systems, the shutdown function may not be as robust. All experimental results depend on the choice of materials, manufacturer, quality, and design. For example, it was shown [28] that when the battery temperature reaches the separator shrinkage temperature, the battery shorts internally and goes into a thermal runaway situation.

Overcharge abuse tolerance of shutdown separators highly depends on the test conditions. Modest changes in the voltage limits or charge rates can have a large impact on the performance of the separator. Moreover, cell failure can occur at temperatures well below runaway temperatures [29]. These results suggest that failure of the separator may contribute to premature cell failure under certain abuse conditions.

Separator designs that provide shutdown should be considered for a specific battery size, application, and potential abuse situations. Separators that are acceptable for mobile phone batteries may or may not be the correct choice to guarantee safety in an electric vehicle battery. Robust testing is essential to make certain that the separators are selected meet both the performance and safety needs for a given application.

Melt integrity

Melt integrity is the ability to exhibit mechanical strength and robustness and resistance to flow above the shutdown temperature, as characterized by a high rupture temperature. A rupture temperature measures the point at which a separator has lost all of its mechanical integrity to form an insulating barrier between the two electrodes. Battery separators used in lithium-ion batteries should maintain melt integrity after shutdown so that the electrodes do not contact each other and create a short circuit. This avoids thermal runaway, particularly when the battery is exposed to high temperatures. TMA is a useful characterization technique to measure the high-temperature melt integrity of separators.

The technique of TMA measures the dimensional changes of a separator under a constant load while the temperature increases at a linear rate. In this experiment, the separators usually show a certain amount of shrinkage and then start to elongate and finally break. A typical TMA graph for a battery separator is shown in Fig. 3.2 to highlight these effects.

The TMA test uses a small separator sample about 5—10 mm long. The sample is held in mini-Instron type grips. The separator is generally held with a constant 2-g load while the temperature is increased at a rate of 5°C/min.

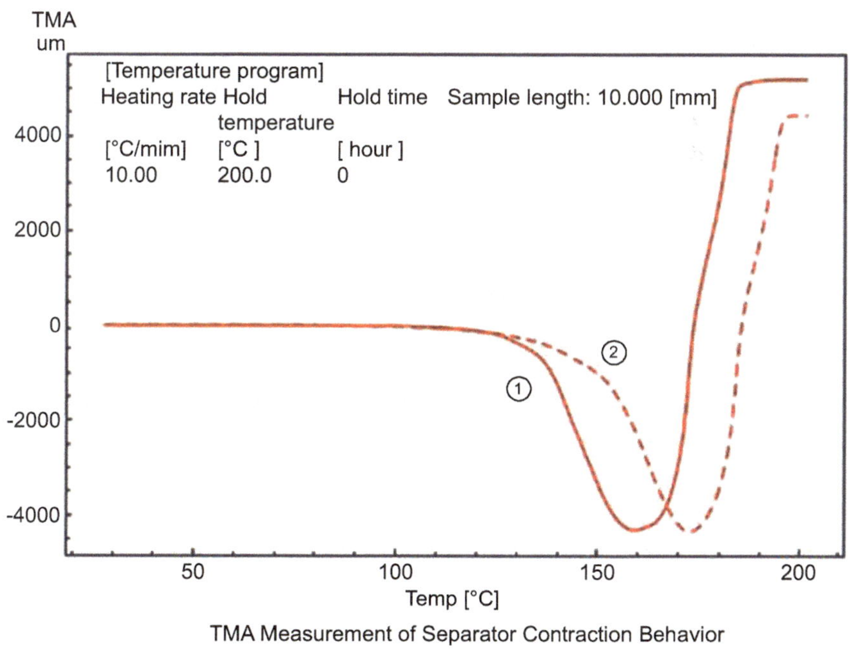

Figure 3.2 Thermomechanical analysis (TMA) data for a battery separator.

The heating occurs until the temperature is reached at which the tension completely ruptures the film. This temperature is usually above the separator material melting point.

Three parameters are usually reported from the TMA test: the shrinkage onset temperature, melt temperature, and melt rupture temperature. This TMA test was shown in the literature to be a reproducible measure of the melt integrity of a separator [30].

Separators with melt integrity values greater than 150°C are desirable for lithium-ion batteries. These values are generally greater than those that can be provided by single-layer PE separators. This is particularly significant for larger lithium-ion batteries used in hybrid and electric vehicles.

The TMA test that has been described is not universally accepted as the best measure of high-temperature melt integrity. Work has been done to develop a standard testing protocol to evaluate high-temperature melt integrity properties in lithium-ion battery separators [31]. The primary rationale for this work was that no standard testing existed to examine high-temperature melt integrity behavior. A further goal of the work was to develop materials that would be used to confirm the validity of the proposed test.

Several tests are described in this work as indicators of the high-temperature melt integrity of the separators. First, a hot tip test is described. In that test, a hot soldering iron tip contacts the separator and the resulting hole is measured. The test measures the dimensional stability of the separator under point heating conditions.

The second test that is discussed is the temperature-dependent ER. The basic features of this test were discussed in the section on shutdown. The ER that is measured can be translated into z-direction stability. The criterion for passing this test is that the separator needs to maintain resistance to greater than 200°C without shorting.

The final test used in this work was the TMA test that was previously highlighted. In that test, dimensional stability under load and temperature was evaluated. The criterion for passing the test was that the separator needed to maintain dimensional stability at temperatures above 200°C.

A correlation was established between the TMA results and the temperature-dependent resistance meltdown to hotbox cell test results. On the other, the hot tip results were less related to the hotbox cell test results than were the other two tests. Lower shrinkage results led to higher heat resistance for the separators. In addition, a formulation was defined that has balanced thermal properties in all three directions. That formulation was based on a coated separator and did not prevent failure of the battery, but it may delay the onset of exothermic events observed in overheating of batteries. That result is typical of the goal of the work to correlate the results of the film tests to the overall performance of batteries.

Future developments in characterization techniques

Most characterization techniques being used examine the separator in isolation and do not consider that its properties may change during the operation of the battery. In fact, the features of the separator in the battery environment are the most important factor that needs to be determined. In that sense, in situ measurements of the separator properties and the changes that occur in them need to be the focus of future characterization work.

Peabody and Arnold [32] simulated normal operation and storage conditions such as external stresses or electrode expansion by subjecting separators to mechanical stimuli. The presence of these stimuli has been shown to lead to separator deformation resulting in increased internal resistance and significant capacity fade in the battery. This capacity fade is a result of viscoelastic creep in the separator that reduces ion transport via a pore closure mechanism. In this work, compressive stress was applied to the battery structure using an Instron testing system. That compressive strain was meant to mimic effects that can occur when the battery is actively charged or stored in a charged state. The work identified an important link between the mechanical properties of the separator and the overall electrochemical performance of the system. By modifying the separator properties, it is suggested that a pathway for improved battery performance can be obtained.

Further work [33] investigated the effects of electrolyte solvents on the mechanical properties of a polypropylene separator. Certain electrolyte solvents such as dimethyl carbonate, diethyl carbonate, and ethyl acetate were found to cause noticeable softening of the separator. On the other hand, in other solvent environments such as propylene carbonate and acetonitrile, the separator retains the mechanical properties of the dry material. This is another example of the need to develop further characterization techniques and methods that address separator properties in the actual environment of the operating battery. Studies such as those in Gor et al. [33] can be used to screen and identify mechanically stable polymer—electrolyte solvent pairs effectively for use in lithium-ion batteries for designed for long life.

An additional advance was made in the development of in situ characterization of battery separators in US Patent 9274068 [34]. That work used a method to detect the separator relative to the electrodes in a lithium-ion battery. The method involves the use of an X-ray—sensitive separator located between the two electrodes. The X-ray—sensitive material in the separator formulation is an X-ray—detectable material; examples of this are metal oxides, metal phosphates, and metal carbonates. The X-ray—sensitive material is generally used at less than 5% by weight of the battery separator. Presumably, this limit is used so that the presence of the X-ray—sensitive material will not negatively affect the mechanical properties and permeability features of the separator.

In actual operation, the battery or battery assembly that contains the X-ray—sensitive battery separator is subjected to X-ray radiation that facilitates detection of the X-ray—sensitive battery separator within the battery assembly. This is important because during winding or battery assembly, the separator portion that extends beyond the lateral edges of the electrodes can be removed, shifted, pushed back, or otherwise misplaced. Such a situation would allow for the possibility of physical contact of the electrodes. An X-ray examination of the assembled battery allows for a check, inspection, or test to determine that the separator remains in position throughout manufacturing. The separator can be observed via X-ray examination to guarantee that it has maintained its position. Most important, the X-ray inspection could be automated by computer to increase the speed of inspection.

There is a continuing need to develop characterization methods such as those that can define the battery structure in an automated fashion. Such characterization techniques can quickly identify potential problems with the batteries and prevent operating issues. As such, these methods can be effectively used in quality control environments to eliminate batteries that could potentially malfunction during operation.

It is highly desirable for these automated techniques to be used to measure and monitor the properties of the separator during actual battery operation. Often if a catastrophic event occurs in the battery, the destroyed battery and corresponding separator are analyzed and tested. An issue with that approach is that is whether the state of the battery separator in the damaged cell existed in the battery before the event or whether the event itself caused additional damage to the separator. Usually, testing is done by disassembling batteries that have undergone catastrophic events and examining the separator from the battery that has been taken apart.

A concern with that approach is that the disassembly process of the battery as well as removal of the separator itself may affect the properties of the battery separator. It would be preferable to examine the separator in the battery itself and see what changes have occurred without completely separating the battery into its component parts through disassembly. In addition, it is highly desirable to measure and monitor the properties of the separator when it is first inserted into the battery between the two electrodes. Certainly, the compressive forces associated with winding and battery fabrication can affect the properties and performance of the separator. Work that has been done using X-ray detection techniques to locate the presence of the separator would seem a good first step toward automatically monitoring and measuring properties of the separator during the operation of the battery. To accomplish that task, a combination of the X-ray detection technique with other characterization techniques will be necessary. Through that approach, it will be possible to follow the development of separator properties as a function of battery operation and define the effect of the observed phenomenon on battery performance.

Overall testing of the stability of the lithium-ion battery separators under potentially extreme battery operating conditions will require leading-edge characterization techniques to determine whether the separators meet performance requirements for large-scale applications. Sometimes, those techniques do not exist and more work is required to develop appropriate tests. Tests that are available for separators used in consumer electronic applications will likely need to be developed for emerging applications. In some cases, totally new techniques and tests will need to be established. The primary point of all tests is that they need to provide information about the separator's performance in the actual battery operating environment. Such information will be the most useful for continuing to understand the features of separators that affect battery performance and how they can be tailored to design next-generation lithium-ion battery separators.

The other area in which work needs to continue is in developing additional characterization techniques that relate separator properties to battery performance. In reality, the separator is one component in the lithium-ion battery that affects battery performance. However, the performance of the battery in actual operation is the final critical output of most interest. All separator properties should be determined in relation to the performance of the battery.

Following up on that idea, work also needs to continue in defining characterization techniques that measure the separator properties with absorbed electrolyte material. It was shown that certain electrolyte materials act as plasticizers for battery separators. The physical properties of the separator that should be considered are the properties of the separator–electrolyte combination. This is because it is the combination that exists in the battery during operation. Hence, those mechanical properties are the ones that are affected during reactions that occur in the lithium-ion battery.

Much work has been done to define appropriate characterization techniques for battery separators. Often, ASTM methods can be used to define the separator features. However, such a standard test does not exist for the property of high-temperature melt integrity. There continues to be a need for such a test, and work needs to continue on defining it.

As new applications for lithium-ion batteries continue to emerge, it is likely that there will also be a need to test and characterize those batteries. As part of that overall characterization effort, tests that monitor the features of the separator will need to be developed. Then, the appropriate correlations of separator properties with battery performance will need to be established. It is only through these developments and the established correlations that the uses of the separator, both current and new, can be completely optimized in emerging uses such as electric vehicles and power grids.

References

[1] ASTM D5947 - 96, Standard Test Methods for Physical Dimensions of Solid Plastics Specimens, ASTM International.
[2] ASTM D - 2873,Standard Test Methods for Interior Porosity of Poly (Vinyl Chloride) (PVC) Resins by Mercury Intrusion Porosimetry, ASTM International.
[3] S. Lowell, E. Shield, Powder Surface Area and Porosity, third ed., Chapman and Hall, New York, 1991.
[4] PMI Conference 2000 Proceedings,PMI Short Course, Ithaca, NY, Oct. 16–19, 2000.
[5] Porous Materials, Inc., http://www.pmiapp.com.
[6] A. Jena, K. Gupta, Presented at the 14th Annual Technical Conference, the American Filtration & Separation Society, May 1–4, 2001, Tampa, Florida.
[7] D. Rajan Rajagopalan Kannan, P.K. Terala, P.L. Moss, M.H. Weatherspoon, Int. J. Electrochem. (2018). ID # 1925708.
[8] C.T. Love, J. Power Sources 196 (2011) 2905–2912.
[9] T. Fujii, T. Mochizuki, U.S. Patent 5,759.678, 1998.
[10] ASTM D88 - 00, Standard Test Method for Tensile Properties of Thin Plastic Sheeting, ASTM International.
[11] C. Xue, D. Jin, H. Nan, H. Wei, H. Chen, C. Zhang, S. Xu, J. Power Sources 449 (2020) 227548.
[12] Y. Xie, H. Zou, H.F. Xiang, R. Xia, J. Memb. Sci. 503 (2015).
[13] ASTM D726 - B, Standard Test Method for Resistance of Nonporous Paper to Passage of Air, ASTM International.

[14] R.W. Callahan, K.V. Nguyen, J.G. McLean, J. Propost, D.K. Hoffman, in: Proceedings of the 10th International Seminar on Primary and Secondary Battery Technology and Application, Florida Educational Seminar, Inc., Fort Lauderdale, FL, 1993.
[15] K.M. Abrams, Electrochim. Acta 38 (1993) 1233.
[16] S. Falk, A.J. Salkind, Alkaline Storage Batteries, John Wiley, New York, 1969.
[17] R.G. Robinson, R.L. Walker, in: D.H. Collins (Ed.), Batteries, The MacMillan Company, New York, 1963, p. 15.
[18] R.B. MacMullin, G.A. Muccini, AIChE J. 2 (1956) 393.
[19] J. Cannarella, C.B. Arnold, J. Power Sources 226 (2013) 149–155.
[20] I.V. Thorat, D.E. Stephenson, N.A. Zacharias, K. Zaghib, J.N. Harb, D.R. Wheeler, J. Power Sources 188 (2009) 592–600.
[21] J.W. Schell, Z. Zhangm, in: The Fourteenth Annual Battery Conference on Applications and Advances, Long Beach, CA, IEEE, New York, 1999, p. 61.
[22] Optodot Corporation SBIR, Shutdown Separators for Protection of Lithium - Ion Batteries Against Thermal Runaway, 2005.
[23] F.C. Laman, M.A. Gee, J.J. Donovan, J. Electrochem. Soc. 140 (1993) 156.
[24] F.C. Laman, Y. Sakitai, T. Hirai, J. Yamaki, S. Tobishima, Ext. Abstr. 6th Int. Meet. Lithium Batter. (1992) 298.
[25] B. Xiong, R. Chen, F. Zeng, J. Kang, Y. Men, J. Memb. Sci 545 (2018) 213–220.
[26] R. Spotnitz, M.W. Ferebee, in: Meeting Abstracts, the Electrochemical Society, Inc., Volume 96 -2 Fall Meeting, San Antonio, TX, October 6–11, The Electrochemical Society, New York, 1996.
[27] W. Matthias, B. Dieter, R. Heinrich, B. Thomas, U.S. Patent 6,511,517, 2003.
[28] E.P. Roth, D.H. Doughty, D.L. Pile, J. Power Sources 174 (2007) 579.
[29] C.J. Orendorff, Electrochem. Soc. Interface 21 (2012) 61.
[30] R. Spotnitz, M. Ferebee, R.W. Callahan, K. Nguyen, W.C. Yu, M. Geiger, C. Dwiggens, H. Fischer, D. Hoffman, in: Proceedings of the 12th International Seminar on Primary and Secondary Battery Technology and Applications, Fort Lauderdale, Florida Educational Seminar, Inc., Fort Lauderdale, FL, 1995.
[31] USABC Battery Separator Development Program, Project ID #ES007, Celgard LLC.
[32] C. Peabody, C.B. Arnold, J. Power Sources 196 (2011) 8147–8153.
[33] G.Y. Gor, J. Cannarella, C. Leng, A. Vishnyakovand, C.B. Arnold, J. Power Sources 294 (2015) 167–172.
[34] U.S. Patent 9,274,068, Assigned to Celgard, LLC, March 1, 2016.

CHAPTER 4

Dry process for battery separator production

Introduction to battery separators

There are two primary functions of a separator in a lithium-ion battery. First, the separator keeps the two electrodes (an anode and cathode) separated during the battery's operation. This assists in preventing short circuits that can occur while the battery is functioning.

The second purpose of the battery separator is to allow ionic transport between the anode and cathode. This transport is essential for effective operation of the battery. To allow this movement of ions, the separator needs to be porous.

Factors for choosing an appropriate separator include good electric insulation with minimal electrolytic ionic resistance. Sufficient physical strength to allow for easy handling, mechanical, and dimensional stability is needed. Chemical resistance to degradation by electrolyte impurities and electrode reactants and products is another requirement. The separator must also be effective in preventing the migration of particles or soluble species between electrodes. Finally, it needs to be readily wettable by electrolyte materials while having uniform thickness.

Two primary production methods are used commercially to make battery separators: the wet process and dry process. The two processes differ in the steps used to produce porous films. Also, the final properties of the separators that are produced by the two approaches are different.

Several excellent reviews exist of the entire battery separator area [1,2]. Those reviews presented summaries of some key features of the separators. Also, comparisons were provided of properties of membranes produced by the different methods. As such, those reviews attempted to cover the entire battery separator field.

Polymer-Based Separators for Lithium-Ion Batteries
ISBN 978-0-12-820120-6
https://doi.org/10.1016/B978-0-12-820120-6.00009-X

There has not been an extensive review of the dry process production method. Owing to increased levels of interest in lithium-ion batteries, particularly for electric vehicle applications, a summary dedicated entirely to the dry process is in order. This chapter provides such a review. A summary of both patent and open literature will be provided.

Several companies produce separators using dry process technology. Celgard is part of Asahi Kasei and is headquartered in Charlotte, North Carolina. It has been among the leading suppliers of battery separators for use in lithium-ion batteries for many years. In fact, the original patents for the dry process separator production were issued to Celanese, previously Ube Industries, Ltd., a key player in dry process separator production. In 2019, it completed the transfer of its separator business to its consolidated subsidiary, Ube Maxell Co., Ltd., which now operates the business. Ube Maxell is a joint venture between Ube Industries and Maxell Holdings, Ltd. With reorganization and integration, Ube Maxell has an integrated framework for production and supply starting from the separator base films. Ube Maxell leverages its product portfolio to meet a wide range of market needs efficiently.

Shenzhen Senior Technology Material sold 120 million m^2 of dry process separators in 2016. That made it the largest dry process separator manufacturer in China at that time. The South China manufacturing base entered production in Nov. 2010. The company led the manufacture of dry process lithium-ion battery separators in China.

Several other Chinese companies developed a presence in the production of dry process separators, including Cangzhou Mingzhu Plastic and Xinxiang Zhongke Science and Technology. Numerous startup companies, particularly in the Asia-Pacific region, are working on producing dry process separators. The sales of these other companies are relatively small compared with those of the other companies.

This chapter begins with a summary of the various steps in the dry process. Particular attention will be paid to process parameters involved in each step and the range of acceptable parameter values. The next section of the chapter summarizes the properties of separators that are produced by the dry process. Then, a discussion is provided of polymers that can be used in the dry process. The chapter concludes with comments about future directions for the dry process, particularly in relation to trends observed for electric vehicles.

Steps in dry process

There are four basic steps in dry process separator production: extrusion, annealing, stretching, and heat setting. In melt extrusion, the resulting lamellar morphology is influenced by the level of stress applied to the molten polymer during extrusion. High extrusion stress levels produce planar stacked lamellar morphologies, whereas low-stress conditions produce a more twisted row–nucleated structure. Further discussion is provided on this point in a later section in the chapter.

Extrusion

Most dry process battery separators are made by a blown film process in which a polymer melt is extruded from an annular die and then collapsed or flattened using nip rolls or the like. As the tubular film is formed, a fluid such as air is continuously blown onto the tubular film. This causes the film to cool and solidify.

It is also possible to produce battery separators using the cast process. In that method, it is important for the chilling or cooling roll to be as close as possible to the extruder, ensuring rapid quenching of the melt. This allows for the retention of a high level of elasticity and drawdown in the precursor film.

Those factors are especially important to the development and formation of a crystalline row structure in which the lamellae are arranged in rows with the long axis perpendicular to the takeoff direction. Such a crystalline structure is desired for the formation of micropores in subsequent annealing and stretching steps in the separator production process. It has been generally thought that the expansion of a blown film in the transverse direction (TD) would disrupt such a crystalline structure and interfere with the formation of micropores.

For polypropylene (PP), a typical extrusion temperature is 220°C, but temperatures of 220–240°C can be used. A cooling roll temperature of 70°C is preferred. Takeup speeds on the order of 50 m/min are used. Drawdown ratios of about 70 m/min during extrusion are usually employed in precursor film production. The necessary extrusion process parameters depend on which polymer is used.

Annealing

The precursor film described in the previous step is annealed to form a well oriented crystalline structure with lamellae arranged in rows with the long axis perpendicular to the machine direction (MD). Such structures are necessary for the formation of micropores because only the stacked lamellae can be stretched to create pores. The annealing process, controlling temperature and time, is carried out at a temperature slightly lower than the melting temperature (T_m) of the polyolefin resin in the film. Annealing for PP is performed at a temperature of about 100°C to about 145°C for 5 to about 30 minutes. Annealing can be done in either a continuous or batch mode. The preferred annealing line speed ranges from 15 to 35 ft/min.

Stretching

Cold stretching

Cold stretching produces cracks on the surface of the polymer sheet. The sheet can be stretched in the MD using stretching rolls. The cold stretching process is performed at a temperature that allows the production of cracks in the amorphous regions of the polymer. The cold stretching process is ideally performed between a glass transition (T_g) temperature of −20 and +70°C. At a temperature that lower than T_g −20°C, high fracture during cold stretching occurs and the formation of uniform cracks is difficult. At a temperature higher than T_g +70°C, the cracks that form are restored by the thermal motion present in the polymer. The stretching ratio in cold stretching needs to be between 10% and 100%. When the stretching ratio is less than 10%, cracks do not adequately form in the amorphous regions of the polymer and the air permeability in the final product is reduced. When the cold stretching ratio is greater than 100%, fracture occurs, which can lead to reduced production efficiency.

Hot stretching

The hot stretching process can be performed in a single step or multiple ones. If multiple stretching steps are used, the first hot stretching step is performed at a T_m of −40 to −10°C. At a temperature than T_m −40°C, fracture may well occur during the expansion of pores in the crack region of the cold-stretched film. The cracks formed through cold stretching are similar to some defects in the polymer, and fracture can occur mainly in the cracks. At hot stretching temperatures higher than T^m −10°C, the pores can be closed because the flowability of the polymer is high. Machine direction stretching at a ratio of 100% to 300% is preferred. If the stretching ratio that is used is too low, the permeability rate of the resulting film may be become too low. On the other hand, if the stretch ratio is too high, the gas permeability rate may become too high (2').

Heat-setting

The purpose of heat-setting is to improve the dimensional stability of the film without adversely affecting the gas flux characteristics of that microporous film. The heat relaxation step is usually accomplished by passing the microporous film under tension over at least two rollers, such that the maximum hot stretched dimension is decreased. The temperature at which the microporous film is heat relaxed is between 25°C below the crystalline T_m of the polymer film and 10°C below the crystalline T_m of the polymer film. For PP, the heat relaxation step is performed using temperatures of 140–145°C. The degree to which the microporous film is decreased in the direction of the stretch is usually about 15%–20%. This is generally accomplished by rotating the downstream roller at a peripheral speed lower than the adjacent upstream roller. Also, the two rollers used for heat-setting can be rotated at the same speed, but the downstream roller will have a greater diameter than the adjacent upstream roller.

Typical separator properties

A typical pore structure for a separator produced by the dry process is shown in Fig. 4.1. The pore shape is that of slit-shaped pores with an aspect ratio generally greater than 5.0. The final porosity depends on the morphology of the precursor film and the specific details of each step in the separator production process. In turn, the structure and orientation of

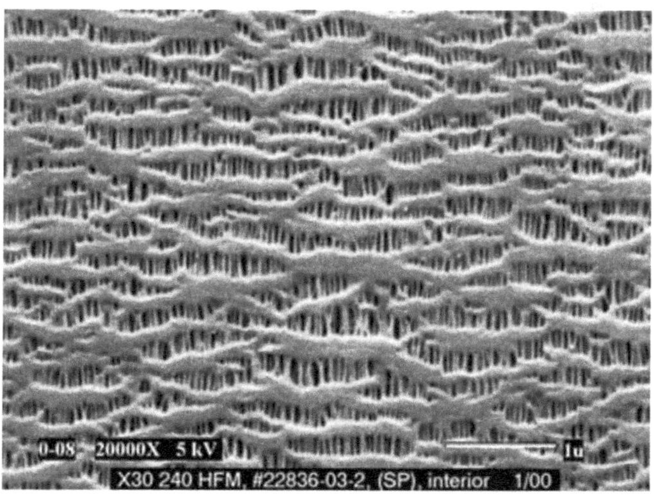

Figure 4.1 Pore structure of dry process separator. Magnification ×20,000.

Table 4.1 Typical property data for dry process separatorsard 2320 Celgard 2400.

Celgard 2320	PP/polyethylene/PP	PP
Thickness, μm	20	25
Porosity, %	41	41
Gurley Japanese industrial standards (JIS)	530	620
Puncture strength, g	360	450

the precursor film depend on the processing conditions used and the polymer resin's characteristics. The pore structure and porosity of the separator are largely responsible for many of the dry process separator properties that will be discussed.

Commercial polyethylene (PE) and PP films prepared by the dry process are available from Celgard LLC [3]. Celgard 2730 and Celgard 2400 are single-layer PE and PP separators, respectively. On the other hand, Celgard 2320 and 2325 are PP/PE/PP trilayer separator materials.

Separators made by the dry process exhibit porosity values in the 40%–50% range. That porosity level can be controlled through the use of appropriate extrusion and processing conditions. Also, the nature of the polymer that is used to produce the separator can have an impact on the final porosity observed in the separator.

Table 4.1 lists key properties of several commercial dry process separators. The ASTM Gurley number is a membrane permeability indicator and is approximately proportional to the square of the tortuosity [4]. The measured Gurley number value is related to both the porosity and the pore dimensions.

A battery separator must be thin to facilitate the battery's energy and power densities. A separator that is too thin can compromise mechanical strength and safety. Thickness should be uniform to support many charging cycles.

The separator must have sufficient pore density to hold liquid electrolyte that enables the movement of ions between electrodes. Excessive porosity hinders the ability of the pores to close, which is a vital function of the separator to shut down an overheated battery. As noted in the data in Table 4.1, a typical lithium-ion separator has porosity values of about 40%.

Celgard 2320 Celgard 2400 material

The puncture strength of a battery separator is used to indicate the tendency of separators to allow short circuits in a cell that may occur owing to holes generated in the separator by the rough surface of an electrode during battery assembly and the charge–discharge cycle. If particulate material from the electrodes penetrates the separator, an electrical short will occur and the battery will be rejected. Generally, puncture strength values should be at least 300 g of force per mil of thickness for lithium-ion cells.

PP, polypropylene.Certainly, other properties of the dry process separators can be discussed in detail. One of the most interesting is tensile strength in both the MD and TD. Because dry process separators are oriented only in the MD, the strength properties of the separators have a large imbalance in the two directions. This has limited the use of the membranes in certain applications. An attempt to address that concern through the biaxial orientation of dry process separators has appeared in the patent literature [5].

Polymers used in dry process separators

The dry process is suitable only for polymers with high crystallinity levels. These include but are not limited to semicrystalline polyolefins, polyoxymethylene, and poly(4-methyl-1-pentene) (PMP). Most of the published literature has focused on isotactic PP [6] and high-density PE [7]. There have also been studies published on dry process separators that are made from polyoxymethylene [8] and PMP [9].

Along with high crystallinity levels, several other polymer parameters have been proposed as significant in the formation of dry process separators. Johnson [10] suggested the importance of the presence of an α_c transition in polymers as a necessary but not sufficient condition for the production of a microporous separator from a polymer. α_c relaxation is associated with the chain axis translational mobility of nonchemical defects in the crystal lattice of the polymer. It has been shown [10] that for PMP, α_c relaxation is required to obtain a microporous membrane.

The molecular characteristics of the resin that is employed, such as the weight average molecular weight (MW) and molecular weight distribution (MWD), are known to have critical roles in the precursor properties in the dry process [11,12]. For resin characteristics, the effect that variables have on the orientation state of the molecular orientation of the melt, specifically the degree of chain extension before crystallization, is a critical factor influencing the morphology and crystal orientation features. The melt-extruded film morphology and orientation provide a window into the melt-flow behavior before crystallization, whereas the degree of the chain extension in the melt influences the precursor properties.

Blends of two linear PPs with different MWs were prepared to develop microporous membranes through the dry process technology [13]. In that work, the role of high-MW polymer chains on row-nucleated lamellar crystallization was investigated. More pore density and a uniform pore size were observed as the amount of high-MW component increased. Also, an increased interconnectivity was observed with the addition of the high-MW PP. As the high-MW PP was added, a slight reduction in mechanical properties along the MD and TD was observed.

In another blend study, a mixture of two PP resins that were different in molecular structure, one with linear chains and the other with long chain branches, was investigated using dry process technology. The presence of the branched component in the blend significantly affected the row-nucleated lamellar crystalline structure in the precursor films. Blending a small amount of the long chain branched PP improved the orientation of both the crystalline and amorphous phases in the precursor films. In addition, the porosity of the blend membrane showed a significant increase compared with the linear PP membrane.

In a study using a different type of polymer [14], Johnson and Wilkes investigated three different isotactic PMP resins with different MW weights and MWD features. The resin characteristics were a key factor in controlling membrane permeability. In addition, as a consequence of the first-stage extrusion parameters, the melt-extruded precursor morphology and orientation were key factors in controlling membrane permeability.

Only certain polymers can be used in the dry process to produce membranes for use as battery separators. Among the most commonly used polymers are PP and PE. Beyond that, appropriate processing parameters are necessary to yield a microporous membrane. This combination of molecular features and process parameters leads to successful production of the desired product. An understanding of the interplay between these features is essential for the production of a battery separator using dry process technology.

Two primary separator structures are produced using the dry process approach [9]: a monolayer PP separator and a three-layer or trilayer separator consisting of a PP/PE/PP structure. In the trilayer separator structure, the PE component provides a shutdown and the PP component gives the separator strength features. Because the PE displays a melting point of about

135°C [15], shutdown occurs in such separators starting at about 130°C [16]. This temperature can be modified through the use of additives [17].

Celgard has been a leader in the development and implementation of dry process technology. They produce and sell several PP monolayer separators as well as different trilayer products. Among the PP monolayer separators are Celgard 2400 and Celgard 2500 separators. The most commonly quoted trilayer product is Celgard 2320.

As stated, when a critical temperature of about 130°C is reached, the porous PE layer softens, collapsing the film pores and preventing ionic conductivity. At the same time, the PP layers provide mechanical support. However, when the internal cell temperature rises to the softening temperature of the separator, the separator has a tendency to shrink because of residual stresses during stretching of the separator films and the difference in density between the crystalline and amorphous phases of the separator materials [18]. In a typical trilayer structure, there is a buffer of only 35°C between the melting point of PE at 130°C and the melting point of PP. If the battery temperature continues to increase after the shutdown event, the separator can fail, exposing the electrodes to internal shorting [19]. In some reported cases, batteries with a shutdown separator remain shut down for as short as 3 minutes before failing owing to internal shorting [20].

The production of trilayer shutdown battery separators using the dry process is described in US Patent 5691077A [21]. In that patent, the PE and PP layers in the separator were subjected to the steps of extrusion, stretching, and annealing previously described as separate films. The individual microporous films were then bonded together, typically using a calendering process, to produce the final battery separator.

Subsequently [22], bonding of the PE and PP layers was performed before stretching in the production process. The final microporous battery separator was thus produced during the stretching of a single precursor material. This approach does not use optimal stretching conditions for the PP layers. This is because the best stretching temperature for the PP is about 140°C, which is above the T_m of the PE layer. Thus, if that stretching temperature were used, the pores in the PE layer would close and the separator would not function effectively. To overcome that issue, the stretching temperature used for this trilayer product is closer to the PE stretching temperature.

Coextrusion has been attempted as an approach to produce a trilayer battery separator using dry process technology [23]. The initial attempts suffered from nonuniform thicknesses in the final film product. However,

Huiqiang produced commercial coextruded products of various thicknesses [24]. They are designated as the HQ series of products. They are touted as having high tensile strength, high puncture strength, and high uniformity. In addition, they possess low heat shrinkage and good high-temperature performance.

While providing a more economic approach to the production of a trilayer battery separator, owing to elimination of the lamination and bonding step, coextruded films have a limitation similar to that of films bonded before stretching. That limitation is based on the fact that optimal annealing and stretching conditions cannot be realized for the PP layers. This is because the optimal annealing and stretching temperature for the PP layer is above the T_m of the PE layer.

In addition to these issues, the production of coextruded film structures can have an effect known as interfacial instability. This issue is related to mismatching of the rheological properties of the polymers used in the coextruded film structure. Particularly significant can be differences in the melt viscosities of the two polymers at the processing temperature that is used. These differences can lead to various visual defects in the film structure, such as the appearance of zigzag patterns in the extruded film. This issue can be resolved by matching the viscosities of the two polymers in the different layers in the film. However, these restrictions further limit the types of polymers that can be employed to produce acceptable structures for use in the dry process technology. These restrictions are in addition to limitations of the general class of polymers that can generate the appropriate morphology to produce pores with the dry process approach.

Separators made by the dry process exhibit significantly lower tortuousities than those made by other processes. Typically, separators made by dry process technology are better suited for high-power density applications. This is because straight pores in separators that are made from the dry process allow for a more direct path for ions to travel, which leads to the attainment of higher power densities in the lithium-ion battery. In general, the tortuosity of the separator can have a vital role in controlling and inhibiting dendrite growth, a factor that can lead to short circuits occurring in the battery [25]. The high tortuosity in a microporous separator may also have a critical role in improving the life cycle of certain lithium-ion batteries. This is because the ionic mobility in the separator membrane depends on the tortuosity of the pathway for ion transport [26,27].

Separators used in lithium-ion batteries are exposed to oxidizing conditions on the side that faces the positive electrode. Also, the separators experience chemical reducing environments on the side that faces the negative electrode. The separators themselves need to be stable in the battery operating conditions during long-term cycling, especially at elevated temperatures. Separators that possess poor oxidation resistance can lead to poor high-temperature storage performance and poor long-term cycling behavior of the batteries. The oxidation resistance of trilayer separators made by the dry process with PP as the outside layer is excellent. This is because of the oxidation resistance properties of the PP that is in contact with the positive electrode in lithium-ion batteries.

From an economic standpoint, dry process separators are generally less expensive to produce than are separators manufactured using other production technologies. This is primarily owing to the smaller number of steps used in the dry process compared with other separator manufacturing processes. Also, the dry process approach does not use additives in addition to the base resin. The material costs to use the dry process are less than for other technologies that require the use of costly additives for implementation.

Along with the advantages of the dry process that have been discussed, there are certain limitations to the dry process technology. First, as was briefly discussed, the dry process is suitable only for polymers with high crystallinity. This limits the type of separators that can be produced in terms of properties. However, blends of immiscible polymers in which at least one polymer has a crystalline structure can be used. Examples of such blends are PE/PP, polystyrene/PP, and poly(ethylene terephthalate/PP blends.

Separators that are produced by the dry process have unbalanced mechanical properties in the MD and TD. Because the dry process separators are oriented only in the MD, the mechanical properties in that direction are higher than in the TD. This means that tearing or failure of the separator is more likely to occur in the TD than in the MD. Processing of the separator that involves the TD must be done carefully to avoid tearing because of the low mechanical properties.

The final concern with the current dry process technology is that it is done in a batch mode rather than a continuous operation. Each step in the production process is done independently of the other steps and the film is physically moved from one piece of equipment to another. The overall final yield for the separator production is the product of the yields for each individual step. This translates into the yield being on the order of

40%–45% [28] for the overall process. This overall yield would likely be increased if the dry process were operated in more of a continuous process than is currently done.

Future developments

There is a push in the battery separator field to produce increasingly thinner separators. This is being driven by the overall desire to incorporate more active electrode material into the battery and thus to increase the power and energy densities of the battery. Such batteries are being used especially in electric vehicles.

As the separator becomes increasingly thinner, the mechanical properties, particularly the puncture strength of the separator, also decreases. This is potentially a concern for safety considerations of the battery, because the puncture strength of the separator is a measure of the safety that it provides to the battery. Finding ways to increase the puncture strength of the dry process separators are an approach to address this concern.

Celgard has addressed this issue in a patent [29] that discloses the use of a high-MW, low-melt flow index PP resin in the layers of a trilayer separator. The use of such a PP leads to an increase in the puncture strength across the 12 to 30-μm range of thickness of the separators. This is described as the membrane's being better able to withstand the rigors of the winding process of the battery assembly process.

Celgard also produced hybrid membranes that combine one layer of a membrane manufactured by a dry process with one layer of a membrane manufactured by a wet process (details are described in Chapter 5). The final separator was produced by laminating wet process PE and dry PP of different thicknesses. As much as a twofold improvement in puncture strength was observed using this approach.

As discussed, another area in which work continues in dry process technology is the development of a coextrusion process to produce trilayer separators. Such an approach would eliminate the need to have a bonding or lamination step in the manufactures of the three-layer separator. As such, the manufacturing costs of this process would be lower than for current trilayer production. This could translate into a lower selling price for the resultant separators.

However, coextrusion brings with it additional processing considerations beyond the production of single-layer films [30]. For coextruded films, it is important to know a great deal about material rheology and compatibility, as well as die design, This is in addition to general requirements that the polymers must satisfy for use in dry process technology. These additional requirements for the coextrusion process make the selection of resins even more critical when coextruded films are produced.

The structure—processing—property relationships of the polymers need to be considered. The most important rheological parameter to consider is shear viscosity. The most important factors that affect polymer viscosity are MW, MWD, and long chain branching architecture. It is crucial to obtain specific shear viscosity curves for each resin.

Three important parameters need to be considered in the die design process: pressure, shear rates, and shear stresses. If the pressure is too high, there will be limits to the operating window in terms of polymer output. On the other hand, too low a shear rate or shear stress increases the residence time of the polymer, which can lead to premature polymer degradation.

Viscosity versus shear rate curves of the different polymers in the coextrusion process can be helpful to factor in pressures and shear rates in the extrusion process. This will allow for the best die design to be achieved. Ideally, the viscosity versus shear rate curves should be generated at the temperature ranges for which coextrusion will be performed.

With the added benefits of coextrusion, there are challenges to be considered and addressed. One of these issues is the potential for flow instabilities. Two of the most common flow instabilities that are observed are the zigzag and wave types. Zigzag instabilities are often commonly observed between two materials that have vastly different shear viscosities. On the other hand, wave interfacial instability is commonly seen when two materials have different elongational viscosities.

The elongational viscosity of a material is a description of the ability of the melt to stretch and extend. It is a measure that influences the polymer's melt strength. Materials that show shear-thinning rheological properties can possess different elongational features. Not only the shear viscosities at high shear rates, but also shear viscosities at low shear rates can have a large influence on the elongational viscosities of a material.

Two features of the materials are extremely important for the production of an appropriate coextruded film product. First, the viscosity of the two polymers should be matched. At the interface of the film, there is a continuity of shear stress values. Because of this, there is an optimal range of viscosity ratios as a function of layer ratios that allow the system to be stable.

Second, to address the issue of elongational viscosity, the average velocity of the two coextruded layers should be matched. This is a relatively simple approach to determine whether flow instabilities could exist between the layers. This could be used because elongational viscosities can be difficult to measure experimentally.

Thus, coextrusion offers many benefits for dry process separator production. However, it also creates some additional challenges. A wide variety of material combinations and equipment designs are available. When you consider the use of coextrusion for dry process separator technology, though, there is an additional limitation on the material and equipment available. These additional demands require a good understanding of factors that can influence the success or failure of the process. Two main areas must be considered for the further development of the coextrusion of battery separators using the dry process. First, the structure—processing—property relationships of the various materials must be well-understood. Second, the optimum extruder—die configurations to produce dry process battery separators must be achieved. It is possible that appropriate equipment for other applications may not be optimal for dry process separator production.

The other area in which work needs to be done on dry process battery separator technology is in the development of additional processing schemes that allow for more types of polymers to be used. Most commercial work has been done on PE and PP. However, the original work [31] on the dry process cited many different polymers as being usable. Work on developing the appropriate process for these different polymers has been relatively limited [8—10]. An interesting development in that regard was made in Wei and Haire [5]. In that work, process conditions were defined that allow for the biaxial orientation of membranes made using the dry process. Implementation of the technology in that patent permits the development of membranes with unique properties. Use of the technology has also allowed the production of membranes from materials that had not been used previously in the dry process [32]. Additional work focusing on developing unique processes to produce battery separators from the dry process is warranted.

Thus, it is apparent that a complete understanding of the structure—property—processing relationships of the polymers used in the dry process technology is essential to further developments. This is true of the current process technology, but it is also critical to developing technologies such as coextrusion. Previously, work did not always focus on such material features; instead, it often used established approaches to deal with new materials.

This approach partially ignores the unique features of new materials and how those novel features can affect the processing characteristics of such materials. The adoption of novel processing schemes to new materials could lead to final separators with unique properties for emerging applications.

References

[1] P. Arora, Z. Zhang, Chem. Rev. 104 (2004) 4419.
[2] [2a] S.S. Zhang, J. Power Sources 164 (2007) 351;
[2b] CA02212469, Assigned to Ube Industries, 1997.
[3] R. Callahan, et al., U.S. Patent 6,602,593, 2003.
[4] F.G.B. Ooms, E.M. Kelder, J. Schoonamn, N. Gerrits, J. Smedinga, G. Calis, J. Power Sources 97−98 (2001) 598−601.
[5] X. Wei, C. Haire, U.S. Patent 8,795,565, 2014.
[6] Tabatabera, et al., J. Membr. Sci. 325 (2) (2008) 772−782.
[7] Lee, et al., Polymer 47 (2006) 3540−3547.
[8] M.B. Johnson, G.L. Wilkes, J. Appl. Polym. Sci. 84 (2002) 1762.
[9] Lee, et al., Energy Environ. Sci. 7 (2014) 3857.
[10] M.B. Johnson, Investigations of the Processing - Structure - Property Relationships of Selected Semicrystalline Polymers, Ph.D. Thesis, Polytechnic and State University, Virginia, 2000, 2000 - 09 - 06.
[11] R. Pazur, R. Prudhomme, Macromolecules 29 (1996) 119.
[12] W.F. Maddams, J.E. Preedy, J. Appl. Polym. Sci. 22 (1978) 2721.
[13] S. Tabatabae, P.J. Carreau, A. Ajii, J. Membr. Sci. 325 (2) (2008) 772.
[14] M.B. Johnson, G.L. Wilkes, J. Appl. Polym. Sci. 84 (5) (2002) 1076−1100.
[15] L. Contat - Rodrigo, A. Ribes - Greus, C.T. Imriew, J. Appl. Polym. Sci. 86 (3) (2002) 764.
[16] X. HUang, Solid State Electrochem. 15 (2010) 649.
[17] Shutdown Battery Separator Made with a Blend of Polymer and Oligomer, U.S. Patent6,749,961, Assigned to Celgard, Inc.
[18] S.S. Zhang, J. Power Sources 364 (2007) 351.
[19] P.C. Balakrishnan, R. Ramesh, T. Prem Kumar, J. Power Sources 155 (2006) 401.
[20] E.P. Roth, D.H. Doughty, D.L. Pile, J. Power Sources 174 (2007) 579.
[21] U. S. Patent 5691077A, Shutdwon Trilayer Battery Separator, Assigned to Celgard, Inc.
[22] U.S. Patent 5,952,120, Method of Making a Trilayer Battery Separator, Assigned to Celgard, LLC.
[23] U.S. Patent Application 2008118827A1, Assigned to Celgard, LLC, 2008.
[24] Huiqinag New Energy website. www.huiqiang.com.
[25] U.S. Patent Application 20170222281, Assigned to Celgard, LLC.
[26] N. Epstein, Chem. Eng. Sci. 4463 (1980) 777.
[27] D. Dijian, et al., J. Power Sources 172 (2007) 416.
[28] Research and markets, in: Global Li - Ion Battery Separators Market Analysis & Trends- Industry Forecast to 2027, November 30, 2018.
[29] U.S. Patent 10,333,125B2, Multilayer Microporous Separators for Lithium Ion Secondary Batteries and Related Method, Assigned to Celgard LLC, June 25, 2019.
[30] C. Tzoganaki, J. Perdikoulias, Polym. Eng. Sci. 40 (5) (May 2000) 1056−1064.

[31] H.S. Bierenbaum, R.B. Isaacson, M.L. Druin, S.G. Plovan, Ind. Eng. Chem. Prod. Res. Dev. 13 (1) (1974) 2–9.
[32] U.S. Patent 10,544,521, Copolymer Membrane, Fibers, Products and Methods, Assigned to Celgard, LLC, January 28, 2020.

Further reading

[1] R.S. Baldwin, et al., NASA/TM -2010 -216099, 2010.
[2] F. Sadeghi, A. Ajii, P.J. Carreau, J. Polym. Sci. B Polym. Phys. 148 (2008).
[3] J.Y. Lee, B. Bhattacharya, et al., Electrochim. Acta 54 (18) (2009) 4312–4315.
[4] U.S. Patent 10,601,012, Multilayer Hybrid Battery Separators for Lithium Ion Secondary Batteries and Methods of Making Same, Assigned to Celgard LLC, March 24, 2020.

CHAPTER 5

Wet process for battery separator production

Introduction

The previous chapter provided many details about a common production process for separators, the dry process. Another process, called the wet process, is used commercially to manufacture lithium–ion battery separators. Details of the wet process and properties of the resultant separators differ considerably from dry process technology. This chapter discusses the wet process in detail and provides information about the typical properties of lithium–ion battery separators that are manufactured through the wet process.

Asahi Kasei is a leading global producers of wet process separators. At the end of fiscal year 2018, their production capacity was 410 million m^2/year for wet process separators. By the end of fiscal year 2021, the wet process production capacity was expected to increase to about 1 billion m^2/year with planned capacity expansions.

Another producer of wet process separators is Toray. The technology is used by Toray to produce separators was initially developed by Tonen Chemical, a subsidiary of Exxon Mobil. In 2019, Toray announced that it would establish a new production facility for wet process battery separator films. It was to be constructed in Hungary and would increase Toray Group's total battery separator film capacity by about 20%. The new facility was scheduled to be operational in Jul. 2021.

There has been much activity in China in the production of wet process separators. In 2016, the sales leader of wet process separators was Shanghai Energy New Materials Technology Co. It sold about 100 million m^2 of products to customers, which included Samsung, LG Chem, and BYD. The top three wet process separator sales in China to Shanghai Energy New Materials Technology Co., Suzhou Green Power New Energy Materials Co., and Foshan Jinhui Hi-Tech Optoelectronic Material Co.

Polymer-Based Separators for Lithium-Ion Batteries
ISBN 978-0-12-820120-6
https://doi.org/10.1016/B978-0-12-820120-6.00008-8

The leading company in the United States that produces separators using the wet process is Entek. The company employs knowledge about battery chemistry, separator properties, and in-application use, which sets it apart from the competition. The technology that is used in their separator production process was initially developed for lead—acid batteries; it was adapted to lithium—ion battery separators. Entek provides a variety of separators ranging in average thickness from 5 to 20 μm.

Steps in wet process technology

The wet process consists of multiple steps: mixing, heating, extruding, stretching, and additive removal. First, polymer resins are mixed with some type of paraffin oil, antioxidant materials, and other additives. The produced mixture is then, heated to produce a homogeneous solution. The heated solution is extruded through a sheet die to produce a gel-like film in the next separator production step. Subsequently, the gel-like sheet is stretched or oriented, usually in two directions. Finally, the additives are removed using a volatile solvent to form the final microporous separator product.

The wet process can be used for both crystalline and amorphous polymers. Often wet process separators use ultrahigh molecular weight PE (UHMWPE). However, other polymers can be used as well.

A detailed discussion of each step in the wet process will not be provided. It should be noted that the wet process is also referred to as the phase inversion process in the literature. As such, studies quoted in this chapter often use the term "phase inversion process" to describe what is called the wet process here [1,2].

Extrusion

First, the mixing process will be discussed. A hydrocarbon liquid or some other low molecular weight substance is mixed with a polyolefin resin. In addition, compositions were developed that contain various fillers in addition to the resin and hydrocarbon liquid [3]. More recently, a variety of inorganic fillers, such as silica and alumina, were added to wet process formulations [4]. Factors such as the surface area and surface chemistry were shown to be key factors in the selection of an appropriate filler for the formulation.

The primary function of the hydrocarbon liquid in the formulation is to act as a processing aid for the polymer resin. In the case of UHMWPE formulations, this is required because the melt viscosity is too high for normal extrusion processing. The processing aid acts as a plasticizer to lower the melt viscosity to facilitate melt processing. Typically, the hydrocarbon liquid is a type of petroleum or lubricating oil. It should show some compatibility with the polymer resin that is being used.

Attention will now be given to the extrusion step in the production process with particular attention to process parameters that are used. The extrusion process is performed at elevated temperatures typically ranging from 180 to 250°C. The actual extrusion temperature used depends on the thermal stability properties of the oil plasticizer used in the formulation. Higher extrusion temperatures will lead to lower melt viscosities of the polymer—oil mixture. However, the extrusion temperature must not high enough to cause chemical degradation of the oil used in the mixture. The degradation temperature of the oil depends on the type of oil that is used in the process.

Another important feature in selecting the extrusion temperature is the desire to produce a miscible mixture of the resin with the plasticizer. This usually leads to the use of increasingly higher extrusion temperatures, because higher temperatures normally lead to increased solubility [5]. However, the thermal stability features of the solvent ultimately define the highest processing temperature that can be employed during the extrusion process.

In WO 2008/016714 [6], some typical extrusion conditions are provided. The melt blending temperature can range from 140°C to about 250°C, or from about 170°C to about 240°C. That research says that when the extruder is a twin screw extruder, the screw parameters are not critical. For example, the screw can be characterized by a ratio L/D of screw length L to screw diameter D in the twin screw extruder, which can range from about 20 to about 100, or from about 35 to about 70. When L/D is less than about 20, melt blending can be more difficult, and when the L/D is more than 100, faster extruder speeds might be necessary to prevent excessive residence time of the mixture. This can lead to undesirable molecular weight degradation. Although it is not considered a critical parameter, the cylinder of the twin screw extruder can have an inner diameter in the range of about 40 to about 100 mm, for example.

The composition of the resin—solvent mixture depends on the nature of the resin used as well as the desired properties of the final product. A function of the solvent in mixtures with UHMWPE is to reduce the overall melt viscosity so that processing can be done with standard film equipment.

Thus, a sufficient solvent needs to be added to the resin to obtain acceptable melt viscosities. On the other hand, an excessive amount of oil can lead to viscosity features similar to the solvent itself. Such mixtures can be difficult to process into usable film products. The optimum amount of solvent to be used will depend on the actual viscosity features of the resin.

The other factor that defines the amount of solvent to be used is the desired porosity of the final separator. This is because the extraction or removal of the solvent later in the production process leads to the development of porosity in the separator. Thus, the amount of porosity that develops in the separator is related to the amount of solvent used in the resin–solvent mixture. The use of an increasing amount of solvent leads to higher porosity [7]. US Patent 6,245,2727 states that it is possible to control the permeability of the membrane by changing the polyolefin composition concentration. The desired porosity depends on the final end use application for the separator. Of course, higher-porosity products will have decreased mechanical strength, a factor that must be considered when using such membranes.

The most common polymer used in the wet process is polyethylene (PE). Among the various PEs available, UHMWPE is the most common version of PE employed. The molecular weight of UHMWPE is noted as being between 3.5 and 7.5 million atomic mass units [8]. UHMWPE powder materials are available and are produced by Celanese, Mitsui, and Braskem, among others. The different producers manufacture UHMWPE materials that are different in their average molecular weights.

UHMWPE possesses a higher mechanical strength than other PE materials. Also, it has a gel-like structure when it melts. These two features can contribute to separators that can display better safety properties than separators based on other materials.

The PE component in the separator formulation can also be a mixture of PEs of different molecular weights [9]. In that patent, the first PE has a weight average molecular weight of 1 million. The second PE in the mixture has a weight average molecular weight less than 1 million. The preferred composition that is discussed is about 30 wt.% of the first PE and about 70 wt.% of the second PE. A primary purpose of the second polymer is to provide shutdown to the separator, an effect discussed in an earlier chapter. A PE of greater than 1 million does not provide effective shutdown when it is used by itself as the resin component in the formulation. In addition to PEs of different molecular weights being used to provide shutdown, different PE types can be mixed with the UHMWPE to yield shutdown characteristics for the separator. Examples of this are high-density PE and linear low-density PE.

Other materials can be included in the formulations in addition to the resin and solvent. These can be additives to guarantee the thermal stability of the resin, such as antioxidants, or they can be various inorganic filler materials such as alumina, silica, or titanium dioxide. The primary purpose of those inorganic materials is often to improve the temperature performance of the separator. This general concept will be further discussed in Chapter 6 of this book.

The polymer—diluent mixture is extruded through a die to form an extrudate that is then cooled to form a sheet. The cooled sheet has the appearance of a gel and hence is called a gel-like sheet. The polymer—diluent mixture is in the molten state during extrusion and is exposed to temperatures of 140°C to about 250°C during extrusion. The extrudate is usually cooled until it reaches a temperature at which the sheet begins to gel. The cooling rate is typically about 50°C/min. Additional details about appropriate process conditions for performing the extrusion can be found in PCT Publication WO 2007/132442 [10].

Stretching

After extrusion is completed, the next step in the wet separator production process is stretching of the cast sheet [11]. In that step, the gel-like sheet is biaxially stretched while it is heated. Stretching is carried out using a tenter or roll or a combination of the two methods. Stretching is done at a given stretch ratio. It can be done in the machine direction (MD) and transverse direction (TD) either simultaneously or sequentially. US Patent Application 20110117439A1 reports that simultaneous stretching is the preferable stretching method.

For PE—diluent-based gel-like sheets, the stretching temperature to be used is 116 to 125°C, preferably between 116 and 123°C. If the stretching temperature for these sheets is above 125°C, the molecular orientation induced by stretching will not occur because the polymer will melt and soften. Melting of the resin can lead to a decrease in overall membrane porosity. This is because melting of the polymer can lead to closing of the pores in the membrane. If the stretching temperature is below 116°C, the resin is insufficiently softened. The membrane may tend to break during the stretching process and the stretching ratio might not be well-controlled. Furthermore, elongation of the microporous membrane will be insufficient.

Additional information about the stretching process can be found in WO 2010/004918A1 [12]. The extrudate can be stretched simultaneously in the MD and transverse direction (TD) at 4.0–6.0 while exposing the extrudate to a temperature of about 110°C to 120°C with a preferred stretching temperature of 115°C. The MD and TD stretch ratios can be the same, but it is not required. In one particular embodiment, both MD and TD stretching ratios are equal to 5.0.

Ihm and others [2] studied the effect of film drawing conditions on the properties of wet process separators. Those workers used a laboratory biaxial orientation machine to produce final membranes with a thickness of about 25 μm. Draw ratios of 2–5 were examined in that study. The biaxial orientation was performed at temperatures of 125 to 130°C.

The mechanical strength and pore structure are influenced by stretching conditions such as the drawing temperature, drawing speed, and draw ratio. The mechanical strength increases with an increase in the draw ratio. The pore size distribution depends on whether the stretching is done before or after solvent extraction. Asahi Chemical adopts the process of stretching after extraction [13]. On the other hand, Tonen makes separators by solvent extraction after stretching [14]. The separators that are produced by drawing after extraction have much larger pore sizes and wider pore size distributions than those produced by extraction after stretching.

The stretching ratio operates in a multiplicative factor on the film size [15]. For example, a film with an initial width of 2.0 cm that is stretched in the TD to a stretch ratio of fourfold will have a final width of 8.0 cm. That patent teaches that the stretch ratio should be equal to 5 in both the MD and TD. Also, the stretching temperature should be in the range of 112 to 119°C.

Thus, for the stretching conditions, for typical wet process separators based on PE, stretching can be done either sequentially or simultaneously. Stretching ratios of at least 5.0 in both directions are preferred with higher stretching ratios yielding separators with higher mechanical properties, such as strength. The optimum stretching temperature is 110 to 120°C.

Extraction

The next step in the wet process separator production process is extraction of the diluent from the stretched sheet. The stretched sheet is washed with a solvent to remove residual diluent. It is then dried and heat-set. Some solvents used for diluent-removing treatment may be volatile, including hydrocarbons such as pentane, hexane, and heptane. They can also be chlorinated hydrocarbons such as methylene chloride or carbon tetrachloride. Fluorinated hydrocarbons such as trifluoroethane can be used. Finally, ethers such as diethyl ether or dioxane can be used as the solvent in the removal process. The solvents can be used individually or combined. The selection of an appropriate solvent depends on the nature of the diluent that is used in the polyolefin mixture.

The extraction of the stretched sheet with the volatile solvent should be performed to a level in which the content of the residual diluent is less than 1 wt.%. Levels higher than that can cause negative features in the battery that contains the separator [16]. The stretched sheet is finally dried to remove the washing solvent. This is typically done by a heating or air-drying method.

An alternative approach to removing the diluent was developed by Brueckner [17]. The Evapore approach involves evaporating volatiles without the additional need to heat the battery separator. The benefits of the technology include low shrinkage and possible high-porosity products. Downgauging of the membranes without property loss and good thickness tolerances are possible. Finally, film characteristics that are evenly distributed across the film web, including low electrolyte resistance and high puncture strength, are possible with this technology.

Brueckner also provides entire battery separator production lines based on the Evapore process. The line concept includes the full process know-how as well. The process benefits from Brueckner's experience and knowledge in wide and fast film stretching. The initial products were focused on the electric vehicle market.

Heat setting

After the extraction step, the dried sheet is thermally heat-set at temperatures of 115 to 120°C. It is believed that if the temperature in the heat-setting step is higher than that to which the membrane is exposed during stretching, a membrane will be produced that has reduced TD heat shrinkage. If a temperature below 115°C is used, the overall strength of the membrane may be inadequate. On the other hand, at temperatures above 120°C, thermal shrinkage may be excessively high. This will have the effect of decreasing the membrane's elongation. The separator should be heat-set in such a manner as to shrink its width by 10% or less. Such a treatment will improve membrane elongation. Shrinking the width by more than 10%

during heat-setting is undesirable. This is because it lead to possible deterioration of membrane permeability; heat-setting the membrane has a tendency to close some of the pores that are present in the initial membrane.

Certain companies claim to perform extraction and stretching steps in different orders. Thus, Asahi and Mitsui perform solvent extraction first and then stretch the membrane. On the other hand, Tonen performs extraction after the stretching operation is complete. These different orders for the process have the overall effect of providing membranes with different pore sizes. Thus, the process as practiced by Asahi and Mitsui yields membranes with a much larger pore size (0.24−0.34 μm) and wider pore size distribution than products produced by Tonen, which have a pore size of 0.1−0.13 μm [2].

Structure of wet process separators

Separators made by the wet process are shown in Fig. 5.1. The pore structure of all of these membranes is similar. However, membranes from Asahi have significantly larger pores than those from Tonen. This is primarily because of the different order of extraction and stretching steps in membrane production.

The planar surface of the final membrane is composed of a network of microfibrils. The microfibrils are composed of polymers used to produce the membrane. The average microfibril diameter can be measured using

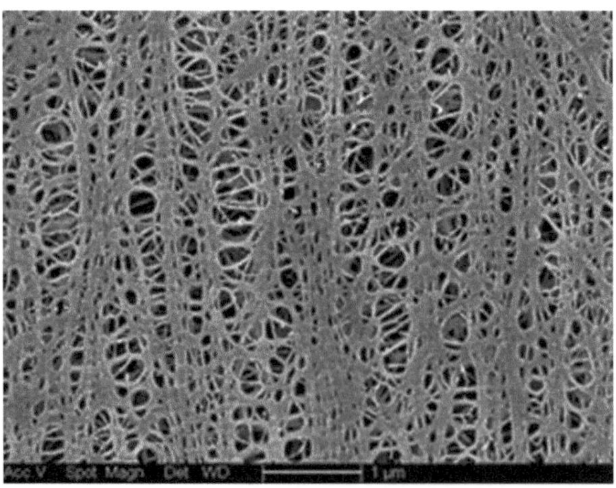

Figure 5.1 Pore structure of wet process separator.

atomic force microscopy (AFM) or scanning electron microscopy (SEM). When AFM is used, the average microfibril diameter and average distance between microfibrils can be obtained from AFM images. This is done by averaging the measured size and spacing of the microfibrils in the images.

For the Tonen samples, the average diameter of the microfibrils is 20–100 nm. The average distance between microfibrils (i.e., adjacent microfibrils) is more than 400 nm. Typically, the average diameter of the microfibrils is 40–70 nm and the average distance between them is 450–650 nm.

SEM can also be used to determine the features of the microfibrils [18]. In that method, a cross-section of the macroporous membrane is observed at a predetermined magnification under conditions of an accelerating voltage of 1.0 kV. An area occupied by microfibrils in the image is determined by arithmetic processing. Then, the total length of the microfibrils is determined. Also, the average diameter between microfibrils can be determined from the SEM image.

As has briefly discussed, the average pore size in wet process separators depends on whether membrane stretching is performed before or after extraction. If stretching is performed before extraction, the average pore size is 0.1–0.13 μm. On the other hand, if stretching is done before extraction, the average pore size is 0.24–0.34 μm. No pores exist larger than 1 μm in either sample.

The PE portion of the wet process separator causes a shutdown beginning at ~135°C, the melting point of PE. However, battery failures in the field demonstrated that a shutdown does not guarantee safety. This is primarily because after shutting down, the internal temperature of the battery is often high enough to cause residual stress and reduced mechanical properties, leading to shrinkage, tearing, or the formation of pinholes.

Properties of wet process separators

Because the pore structure and mechanical strength of wet process separators are not induced mechanically, the properties are not directionally dependent [19]. This means that the separators are nearly isotropic within the film plane. This was further verified and studied using a dynamic mechanical analyzer (DMA) [20].

Some typical properties of wet process separators available from both Asahi and Tonen are provided in Table 5.1.

Table 5.1 Typical properties of wet process separators.

Separator properties	Asahi HiPore	Tonen Sekala
Structure	Single layer	Single layer
Composition	Polyethylene	Polyethylene
Thickness, μm	25	25
Gurley, s	21	26
Porosity (%)	40	41
Melt temperature (°C)	138	137

The table shows that the parameters are similar for the two brands of separators. Moreover, the properties of the membranes can be changed by controlling the composition of the solutions and the evaporation of the solvent in the solidification processes.

Entek provides several different brands of separators produced using wet process technology [21]. The molecular weight distribution of the PE, the percentage and type of plasticizer, extraction and drying conditions, biaxial stretch ratios, and annealing temperatures are all factors that can affect the final structure and properties of the Entek separators. Three products are discussed on the Entek website: Entek EP, Entek EPH, and Entek EPX. Table 5.2 lists typical properties of those separators.

The data show that it is possible to vary the thickness of the separators from 12 to 20 μm. Presumably, even thinner separators are possible. Also, porosity values that range from 43% to 54% are quoted in these data. However, the puncture strength decreases from 550 to 300 g as the separator thickness decreases from 20 to 12 μm. This is a potential issue with the safety of the separator that is provided in the 12-μm case.

Table 5.2 Typical properties of Entek separators.

Property	Units	Entek EP	Entek EPH	Entek EPX
Thickness	μm	20	16	12
Basis weight	g/m^2	10.5	8.4	5.5
Porosity	%	43	47	54
Air permeability	s	380	150	65
Tensile strength	kg/cm^2	MD >1000	MD >1000	MD >1000
		TD 750	TD 700	TD 800
Elongation	%	MD 90	MD 80	MD 90
		TD 300	TD 250	TD 230
Puncture resistance	g	550	430	300
Shrinkage, 120°C, 1 h	%	MD <10	MD <10	MD <12
		TD <3	TD <4	TD <7

MD, machine direction; *TD*, transverse direction.

Entek also investigated the effect of including ceramics in wet process separators [22]. Both silica and aluminum oxide were used in this work at different loading levels. The separators had high-temperature dimensional stability with shrinkage of less than 5% at 180°C. The separators can be produced with high porosity values of 70–80%. They have low impedance values. Rapid wetting with the electrolyte material can be achieved. Two potential issues with the separators is that they have low tensile modulus strength. This is because of the high amount of filler and porosity values that are in the 70–80% range. In addition, the separators do not display shutdown behavior.

In terms of the performance of batteries based on these ceramic-filled separators, high-voltage stability is observed. A silica-filled separator in a battery increases the cell cycle life compared with a control. This is attributed to more uniform electrolyte distribution, which leads to more uniform electrode use. Finally, the cycle life of cells that contain several different inorganic fillers is better than the control. No filler combination is better than precipitated silica in that regard.

In addition to monolayer separators made by the wet process, coextruded separator films were produced [23]. ExxonMobil introduced these materials based on tailored heat-resistant polymer formulations. The multilayer separator materials showed significant improvements in the porosity that leads to lower internal cell resistance as well as thermal properties compared with their predecessors [24]. In this case, improved thermal properties can extend the mechanical integrity of the separator for a long time during a thermal event. This will give the battery an increased amount of time to dissipate heat that is generated, thus decreasing the probability of a more catastrophic event.

The developed film can retain the original shutdown temperature of the polyolefin material and retain its mechanical integrity for an additional 40°C. By not melting until a higher temperature is reached, this separator provides additional time for a malfunctioning cell to dissipate heat, reducing the possibility of transferring the heat to its neighbors and setting off a larger additional thermal event. A larger temperature differential between the shutdown and meltdown temperature is evident in the coextruded material. This translates into a safety enhancement provided by the separator [25].

Mechanical properties of four grades of the coextruded films are reported in R.S. Baldwin et al. The properties are compared with those of a state-of-the-practice separator, Tonen Sekala. It serves as a standard separator component in some lithium–ion battery designs.

The tensile properties in both directions are of a similar magnitude for all film grades. Differences in the tensile properties among the various grades can be attributed to differences in the material composition and film processing. Average shrinkage values for all film grades were higher in the MD than in the TD. Most of the coextruded grade separators had relatively low shrinkage results, as measured by a DMA. All separators had less than 2% shrinkage in the MD and less than 1% in the TD.

Thermal mechanical analysis was used in that work to study the melt integrity of the separators. Three parameters were reported from the testing: the shrinkage onset temperature, deformation temperature, and rupture temperature. For a separator grade with a higher observed deformation temperature, a higher rupture temperature was also observed. The melt integrity data show that the coextruded grades with higher rupture temperatures relative to the control separator would provide higher separator mechanical integrity and electrode isolation for a longer time as the temperature continued to rise to the point of actual rupture of the separator. Furthermore, displacement versus temperature data showed that meltdown of the coextruded separators was more gradual than for the control material.

Regarding the mechanical property data for the coextruded separators, they may provide an enhanced level of internal safety-related performance and reliability for lithium—ion batteries relative to state-of-the-practice separator materials. The improved meltdown features of these separators extend the integrity of the barrier between the two electrodes for a longer time during a thermal event. This allows the battery additional time to dissipate heat and avoid the occurrence of a more catastrophic event. In addition, improved permeability of the coextruded materials may provide enhanced cell-level energy and power capabilities.

In further work [26], separator impedance features as a function of temperature of these materials was examined. A modified 2325 coin cell configuration was used as the cell fixture for conductivity and impedance measurements of electrolyte-wetted separator samples over the desired temperature range. The test configuration was chosen because it provides a constant and reproducible pressure to the separator sample in a sealed cell environment.

The thermal response to the control separator and coextruded battery separator samples was studied by monitoring the relative resistivity and impedance of the electrolyte-wetted separator material while the temperature of the sample was increased at a constant rate. The focus of the study was to evaluate the enhanced high-temperature melt integrity separator properties related to the overall cell-level safety. Two grades of separators that were developed to tailor enhanced thermal stability were studied in this work; the results were compared with a control sample.

The data showed that stable, low cell resistance was observed from room temperature to ~130°C, at which point the PE separator began to melt. As the temperature increased further and the material continued to melt, the pores in the separator continued to close and the cell resistance increased dramatically. As the temperature increased above ~ 150°C and the polymer deformed, flowed, and shrank dimensionally, the observed cell resistance began to decrease because of morphology changes that occurred in the separator. The decrease in the resistance that was observed was likely due to the opening of channels in the material as it deformed. That channel opening can permit the conduction of ions. Between ~170 and 177°C, the rupture temperature for the separator, the cell resistance dropped to zero. This indicates that the separator sample had lost all mechanical integrity to form an insulating barrier between the electrodes. The result of this effect was the occurrence of a short circuit in the cell.

For both coextruded materials studied in this work, the temperatures observed for the onset of the shutdown event were in excellent agreement with values observed from the DMA studies of R.S. Baldwin et al. [23]. The shutdown event was correlated with the melting of the polymeric separator material and the experimental rheological observations were associated with the melting behavior. The response of the resistance observed after the polymer melting temperatures were reached showed that the coextruded materials were able to sustain relatively high resistance values, or mechanical integrity, for a longer time before the onset of the short-circuit event, compared with the control separator material.

The increased thermal stability provided by the coextruded separator materials was further investigated by AFM [27]. Those results showed that when temperatures above the separator shutdown temperature were reached, some of the polymer network appeared to remain unchanged whereas bulk material domains appeared to be embedded within them. These observations suggest that the remaining networks acted as a type of skeleton and absorbed the molten polymer domains into them, maintaining the melt integrity at a high temperature.

The overall results of these electrical studies corroborate the results and conclusions from R.S. Baldwin et al. [23]. The conclusions of both studies were that the coextruded grade material formulations can provide an enhanced level of internal safety-related performance and reliability for lithium–ion batteries compared with monolayer separator component materials. The improved meltdown properties of the coextruded materials, as shown by two different sets of experiments, extend the integrity of the electrode-separating barrier for a longer time during a thermal event. This allows the cell additional time to dissipate heat and avoid the occurrence of an event such as a fire or explosion. The improved permeability features of the coextruded materials may also provide an increased level of energy and power capabilities. To define the potential performance advantages and disadvantages better at the cell level relative to specific separator properties, additional work is necessary on these types of separator materials. That these separators can provide enhanced safety attributes is an important feature of the separator materials for high-power battery development needs for electric vehicle applications.

Toray Tonen also issued a patent on the development of coextruded separators produced using wet process technology [24]. In that patent, die extrusion is described as being conducted using conventional die extrusion equipment. Specifically, two- and three-layer extrudates are discussed. Each layer in the separator is composed of a separate polyolefin solution in that work. The polyolefin solutions are selected independently of each other. The patent discusses the use of polypropylene combined with PE as the resin component of at least one of the polyolefin solutions. In addition, another polyolefin such as poly-4 methyl pentene may be used in one layer of the separator. The produced membranes are touted as having improved permeability, puncture strength, and shutdown speed compared with other membranes based only on PE.

Advances in wet process technology

The initial wet process technology was extended beyond its initial application to monolayer PE separators in which a step in the production process is extraction of the solvent. As briefly discussed, Brueckner modified the solvent extraction process to allow for evaporation of the solvent. This allowed for a reduction in the production costs of the wet process separators to make them more comparable to costs associated with the production of dry process separators. This reduction in cost has made the wet process more economical than previously possible. Additional process enhancements of the wet process should be examined to continue to lower the costs associated with it. For example, work is ongoing to better understand the biaxial stretching process of wet process separators, ways to optimize the stretching process, and the associated parameters to produce the final separators most effectively with the highest yield possible.

The development of coextruded structures and the necessary technology permits the potential incorporation of additional polymers into wet process separators. Such incorporation can allow for the tailoring of properties of the battery separators for specific new applications. This can be achieved by producing mixtures of other polymers with the base PE resin that is used in wet process separators. However, there may be limits to the amount of the second polymer that can be added to the PE in this approach. This is because the other polymer in the mixture often has limited solubility in the same solvent as the PE; thus, mixtures can obtain only a small amount of this second polymer and still maintain solubility of the entire polymer mixture.

In addition to the production of wet process coextruded separators that contain polymers other than PE. work has been done to produce single-layer separators based on polymers other than polyolefins. The wet process, or modification of phase-inversion technology, is applicable to both amorphous and semicrystalline polymers as long as the appropriate solvent system can be found for the particular polymer that is being used. For example, sponge-like polyimide separators were prepared via a soluble precursor and nonsolvent-induced phase separation process [28]. By controlling the process parameters, a polyimide separator with high porosity and pore connectivity was successfully obtained and used in lithium—ion batteries. The separator displayed high thermal stability as well as excellent wettability in the liquid electrolyte and high electrolyte uptake. Compared with cells with usual polypropylene (PP) separators, cells that were assembled with the polyimide separators had lower resistance, higher discharge capacity, and better rate capability. The results suggest that the process technology can be an effective approach to preparing high-performance separators.

In other work [29], high-temperature resistant and fast-wetting lithium—ion battery separators were produced through a phase-inversion process using novel polyetherimides (PEIs). Separators 15—25 μm thick were produced from a PEI based on bisphenol-acetone diphthalic anhydride and para-phenylenediamine. The separators had low degrees of swelling in electrolyte solvents, low contact angles, fast electrolyte wicking, and high electrolyte uptake. The separators that were produced covered a wide range of conductivities. Dynamic mechanical analysis demonstrated dimensional stability up to 220°C. Pouch cells that were assembled using the novel separator had an excellent capacity retention of 89.3% after 1000 cycles. The novel PEI material is a promising candidate for high-performance separators.

A novel approach to producing polyimide separators using the wet process approach was reported [30]. In that work, two porogens, dibutyl phthalate and glycerin, were employed in a nonsolvent-induced phase separation process. Compared with a commercial PE separator, the polyimide separator has significant thermal stability and better ionic conductivity and wettability in both carbonate and ether electrolytes for lithium—ion batteries. Battery coin cells assembled with the polyimide separator still work after heating at 140°C for 1 h.

There was also patent activity in the production of battery separators based on the use of PEIs [31]. That patent uses a phase-inversion process to provide lithium—ion battery separators. The phase separation process can be induced by exposing the polymer solution to a nonsolvent in either the liquid state (liquid-induced phase separation) or the vapor state (vapor-induced phase separation). Key features that included the composition of the polymer solution, coagulation bath, and temperature were studied to obtain the desired morphology and properties in the separator. Both liquid-induced phase separation and vapor-induced phase separation were suitable processes to produce porous separator films with tunable pore structures. These films were suitable for use in battery separator applications.

In the work, separators were prepared that met the usual separator requirements for porosity, pore size, thickness, conductivity, and modulus. The process is versatile in terms of the obtained porosity, pore size, and thickness. This led to separators that were useful in the final performance of an actual electrochemical battery environment. For example, based on the use of novel separators, batteries have an unexpected improvement in life cycle compared with those produced using commercially available polyolefin separators. In addition, the separators possess high-temperature melt integrity as well as electrolyte wettability. Specifically, high-temperature melt integrity values that exceed 180°C are quoted in this patent for separators that were developed.

Conclusions

Details of the wet process technologies for the production of battery separators were presented in this chapter. A specific focus was on an understanding of relationships between the structure of the separator that is produced and the processing parameters that are used in their production. An example of this is that if the order of the stretching and extraction steps in the production process are reversed, a separator material with a different pore size is produced. Certainly, the two different structures will have different properties, and those property differences will likely lead to different performances in the batteries that contain the separator materials.

As efforts continue to produce thinner separators, work will be required to adapt the wet process to meet that need. Two steps in the wet process can be modified to deal with that development. First, films can be produced or cast at a thinner thickness than is currently done. This change may necessitate the need to develop unique die and production equipment. Second, if the cast film thickness remains unchanged, the final film thickness can be modified by increasing the stretch ratios in the stretching step in the production process. If this approach is used, the stretching temperature may need to be slightly changed for to stretch the films effectively to those increased stretch ratios. Also, if an increased stretch ratio is used in the stretching step, it is likely that there will be more breaks in the film during production. This is because thinner separators are weaker than thicker separators, which leads to a higher potential for the film to break during stretching. This affects the overall yield for the entire production process. As such, it can be anticipated that the production costs will increase.

A potential way to offset the decrease in separator strength with a decrease in thickness is to use a polymer that has higher strength. For the wet process, a polymer that can be used in this regard is a UHMWPE with high molecular weight. Such a polymer offers the potential to produce films with higher mechanical properties. However, it is also more difficult to process owing to the increase in melt viscosity that it provides to the polymer solution. In addition, the increased molecular weight will make films harder to orient biaxially to high levels. Thus, there is the typical trade-off between an increase in mechanical properties and the ease of processing of films with this approach.

Another area in which there is interest in wet process films is enhancing the porosity level over values that are currently observed. This interest is driven by the need to manufacture batteries that possess higher energy and power densities than current consumer electronics batteries. One way to achieve that goal is to allow the separator to absorb an increased amount of electrolyte. This can be achieved by increasing the overall porosity of the separator, because the electrolyte is absorbed into the separator pores.

A way to increase the porosity of wet process separators is to increase the amount of diluent or solvent that is used in the polymer solution. This is an effective approach because it is largely the removal of the diluent in the extraction step that leads to the development of pores in the final separator. Thus, it is largely the amount of diluent that is extracted from the separator that determines the level of porosity. Thus, increasing the amount of diluent leads to the production of separators that possess higher porosity levels.

However, the use of high diluent levels in the polymer formulation can lead to issues with the extrusion of the solution into a film. This is because the diluent is a liquid that usually has a low viscosity at typical extrusion temperatures. The presence of the polymer, typically UHMWPE, will increase the viscosity of the extruded mixture. However, because the mixture is largely based on the diluent, the viscosity of the solution will be similar to that of the diluent. This will lead to issues with the production of a film that has sufficient strength to withstand further processing steps. Also, there will be concerns with maintaining a steady flow of the polymer solution to produce an adequate quantity of film for further evaluations. Furthermore, because there is likely to be an uneven flow out of the extrusion die, the film that is produced will need to be examined carefully for nonuniform thicknesses.

If these extrusion issues can be overcome, the film that is produced will need to be handled carefully during additional processing operations. This is because separators that have high porosity values will also have relatively low strength values. These low-strength values will lead to an increased potential for film breaks during stringent processing conditions, such as those that occur during film winding in battery fabrication. In addition, lower puncture strength values of high-porosity films will allow for the facile development of holes in the separators. The presence of those holes increases the potential for short circuits in the battery that can lead to catastrophic thermal events.

As with the development of thinner separators, an approach to dealing with the low strength of high-porosity separators is to use a polymer component with an inherently higher strength than materials that are currently being used. Two examples of such materials are high-crystallinity polypropylene and polyolefins of very high molecular weight. Both of these types of materials have higher mechanical properties than standard polyolefin materials. As such, they can be regarded as special polyolefin grades. These specialty grades often require the use of novel production methodologies and materials, such as metallocene catalysts, for manufacture. Because of this, the cost of these polymers is higher than for standard polyolefin grades, and the higher costs will likely translate into a higher selling price for separators that are produced using these special polymer grades.

Regarding wet process technology, the basics of the necessary process and material parameters were defined for the production of separators used primarily in consumer electronics applications. However, because emerging uses such as electric vehicles are being developed for wet process separators, those processes and materials need to be modified to deal with the needs of these applications. The use of new materials will necessitate the definition and development of new processing parameters to make effective separators. This is an example of the continued need to understand the interplay between the structure and processing relationships of polymers to be able to meet the needs of new applications effectively for lithium–ion batteries. This is a relatively new process for the development of battery separator materials and is an approach that merits further attention and investigation.

In addition to those developments, work needs to continue to reduce the production costs of wet process separators. Some significant developments have been made in this area by Brueckner with the Evapore product, which eliminates the extraction step in the standard wet process. Additional work needs to be done to confirm that the separators produced by this novel approach perform identically to those manufactured using standard wet process technology. The performance must be verified in terms of the properties of the separator itself as well as the operation of batteries that contain the novel separators. Equal or better performance of the batteries is expected; if confirmed, that will completely justify the process modification associated with the production of the Evapore separator. On the other hand, if inferior performance of the batteries is observed, a discussion of the trade-offs between lower-cost products and the required performance of the separators in batteries needs to take place. Thus, the overall potential of the Evapore product to replace current wet process separators in all applications must be assessed. An important question to address in that regard is whether there are certain advantages to using those novel separators in consumer electronics applications that are not possible in electric vehicle applications, or vice versa.

Finally, additional applications for the wet process separators need to be explored. Those applications range from consumer electronics to electric vehicles and power grids. As those new applications are being explored, there will be a realization that a single separator structure is not optimal for all of these applications. Thus, the separator that is the best for use in consumer electronics may not be the best one for use in electric vehicles. However, it may be possible to tailor and fine-tune the properties of a separator that are best for one application to make that separator structure so that it can also be used in other applications. To accomplish that effectively, it is necessary to have a wide knowledge of the structure—property—processing relationships of these wet process separators that are made by the standard process or by novel processing technologies such as those developed by Brueckner for the Evapore product. This should be the general theme of all future developments in wet process separators as new work continues.

References

[1] R.E. Kestig, Synthetic Polymeric Membranes, 2nd. ed., John Wiley & Sons, Inc., New York, 1985 (Chapter 2).
[2] D.W. Ihm, J.G. Noh, J.Y. Kim, J. Power Sources 109 (2002) 388.
[3] A. Michiyuki, Jpn. Patent 8064194, 1996.
[4] R. Waterhouse, J. Emanuel, J. Frenzel, D. Lee, S. Peddini, Y. Patil, G. Fraser - Belland, R.W. Pekala, The 29th International Battery Seminar and Exhibit, March 12—15, 2012.
[5] S. Liu, W. Zhou, W. Yu, J. Membr. Sci. 379 (2011) 268.
[6] T. Kotaro, K. Shintaro, WO 2008/016714, Polyolefin Composition, its Production Method, and a Battery Separator Made Therefrom.
[7] U.S. 6,245,272, Tonen Chemical Company, June 12, 2001.
[8] S. Kurtz, The UHMWPE Handbook: Ultra High Molecular Weight Polyethylene in Total Joint Replacement, Academic Press, 2004.
[9] U.S. Patent Application 20110117439A1, Microporous Membranes and Methods for Producing and Using Such Membranes, Assigned to Toray Tonen Specialty, May 19, 2011.
[10] PCT Publication WO 2007/132442.
[11] U.S. Patent 6,245,272, Polyolefin Microporous Film and Method for Preparing the Same, Tonen Chemical Corporation, June 12, 2001.
[12] WO 2010/004918A1, Microporous Membranes and Methods for Producing and Using Such Membranes, Tonen Chemical Corporation, January 14, 2010.
[13] Asahi — Kasei, Jpn. Patent 8 — 64194.
[14] Tonen, Jpn. Patent 63 — 273651.
[15] U.S. Patent 9,044,716B2, Microporous Membranes and Methods for Producing and Using Such Membranes, Toray Battery Separator Film Co, June 2, 2015.
[16] R. Waterhouse, D. Walker, A. Brown, D. Merritl, J. Emanuel, E. Hostetler, C. La, C. Rogers, R.W. Pekala, Chemical and Physical Changes throughout the Life of a Battery Separator, Entek International, LLC, Sept. 9, 2015.
[17] Plastics Machinery, Feb. 18, 2015, pp. 1—5.
[18] U.S. Patent Applications 2006/0103055, Polyolefin Microporous Membrane, Assigned to Asahi - Kasei, May 18, 2006.
[19] R.S. Baldwin, National Aeronautics and Space Administration, Glenn Research Center, Cleveland, OH, 2009, p. 9.
[20] C.T. Love, J. Power Sources 196 (2011) 2905—2912.
[21] www.entek.com/lithium-ion.

[22] W. Wood, D. Lee, J. Freznzel, D. Wandera, D. Spitz, A. Wimer, R. Waterhouse, R.W. Pekala, Structure – Property Relationships of Ceramic – Modified Separators, March 23, 2016.
[23] R.S. Baldwin, M. Guzik, M. Skierski, NASA/TM – 2011 – 216979/Part 1, 2011.
[24] U.S. Patent 8, 012, 622, Assigned to Toray Tonen Specialty Separator Godo Kaisha, September 6, 2011.
[25] www.greencarcongress.com/2007/05/tonen-liion_bat/html.
[26] R.S. Baldwin, NASA/TM - 2013 - 216979/Part 2, September 2013.
[27] T. Isishara, et al., in: EVS24 International Battery Hybrid and Fuel Cell Electric Vehicle Symposium Proceedings, Stavanger, Norway, 2009.
[28] H. Zhang, C. Lin, M. Zhou, A.E. John, B. Zhu, Electrochim. Acta 187 (2016) 125–133.
[29] R. Abee, F. DaRosa, M.J. Armstrong, M. Hantel, D. Mourzagh, J. Power Sources 345 (2017) 202–211.
[30] M. Li, Z. Zhang, Y. Yin, W. Gao, Y. Bai, F. Zhang, B. Zhao, F. Shen, X. Han, ACS Appl. Mater. Interfaces 12 (3) (2020) 3610–3616.
[31] U.S. Patent Application 20140030608A1, Assigned to SABIC Global Technology, July 29, 2013.

CHAPTER 6

Other separator production processes

Introduction

The last two chapters described the two primary production processes for manufacturing battery separators for use in lithium-ion batteries: the dry and wet process technologies. For a long time, these were the only two primary processes used to make separators. However, as new applications, particularly electric vehicles, developed, new process technologies have been pursued. This chapter will focus on the details of these approaches and how they can produce separators that can meet the challenges offered by emerging applications.

These approaches are largely driven by the electric vehicle market. Thus, an important need being addressed with these separator materials is additional safety that can be provided to the lithium-ion battery. Because of the large size of the batteries that are used, this requirement is more significant than in consumer electronics applications. In addition, the realization that a single separator structure is not optimal for use in all applications has led to an examination of novel production processes.

These processes rely on an understanding of the structure—property—processing relations for polymers that are used to produce battery separators. Those relations will be different from those for polyolefins, polyethylene, and polypropylene (PP); this will affect the performance of the batteries that contain the separators. In addition, the use of novel processes and materials to produce separators will lead to unique separator structures that will need to be handled differently in fabricating the battery. This should not be considered a drawback of these new separators; instead, they should be treated as a potential opportunity to learn about these materials and the processes used in their production.

These novel processes do not simply involve adjusting either the dry or wet process technologies that were highlighted in the previous two

Polymer-Based Separators for Lithium-Ion Batteries
ISBN 978-0-12-820120-6
https://doi.org/10.1016/B978-0-12-820120-6.00003-9

chapters. They are processes that were largely developed for applications other than battery separators. However, to use these processes to produce lithium-ion battery separators, the process parameters need to be appropriately adjusted. Much of the developmental work that occurred in these areas focused on the necessary process features required to manufacture battery separators using the proposed approaches.

Three basic approaches will be discussed in this chapter. The first uses nonwoven materials to produce battery separators. The second technology uses the relatively new method of electrospinning to produce battery separators. The final separator technology that is highlighted is based on the use of a unique additive to produce pores during the biaxial orientation of PP. The biaxial orientation of PP is a well-established technology to produce packaging films. This last technology largely draws on knowledge and expertise that are available on the topic. For each of the three approaches, the technical details of the approach will be described. Then, the advantages and disadvantages of each technology will be discussed. Finally, the way in which each method addresses the needs of emerging applications will be pointed out.

Nonwovens

Several approaches will be discussed to produce lithium-ion battery separators using nonwoven materials. In general, a nonwoven fabric is a fabric-like material made from short staple fibers and continuous long fibers bonded together by chemical, mechanical, heat, or solvent treatment. Some nonwoven materials lack sufficient strength unless they are made dense or reinforced by a backing.

Nonwoven fabrics are flat or tufted porous sheets directly made from separate fibers, molten plastic, or plastic film. They are not made by weaving or knitting and do not need to be converted from fibers to yarn. They are generally engineered fabrics that can be single-use, have a limited life, or be durable. They are used as components of apparel, home furnishings, health care, engineering, and industrial and consumer goods.

There are several methods for producing bonded nonwoven materials, depending on how the fibers are bonded in the final material. Mechanical, chemical, and thermal bonding techniques are used. In mechanical bonding, the fibers are not altered by heat or chemicals; instead, they are bonded by an alternate method such as stitch bonding or needle punch techniques. Chemical bonding, sometimes referred to as adhesive bonding, applies adhesive binders to webs by saturating, coating or spraying, printing, or foaming techniques. Thermal bonding uses heating the web to melt the crystalline region partially, repeat the newly released polymer chain segments across the fiber or fiber interface, and subsequently cool the web to resolidify it and trap chain segments that have diffused across the fiber or fiber interface.

Spun bond nonwovens are made by continuously spinning filament fibers onto a moving belt. The continuous fibers normally come from PP, polyester, or polyethylene raw materials that have been dried, melted, and extruded into a flowing polymer melt. After the spinning process is completed, the polymer emerges as numerous thin, continuous filaments. The filaments are then cooled, stretched, and solidified to consistent dimensions before they are laid on the moving belt to form a mat.

In the wet laid production process, staple fibers up to 12 mm long, often mixed with viscose or wood pulp, are suspended in water using large tanks. After that, the water–fiber or water–pulp dispersion is pumped and continuously deposited onto a forming wire. The water is removed, filtered, and recycled. Besides synthetic fibers, glass ceramic and carbon fibers can be processed using this technology.

Melt-blown nonwoven fabrics are produced by a single step that transforms polymer chips or granules as a raw material into a randomly laid fiber structure. The chips or granules are heated and converted into a polymer melt. The melted polymer is then blown by hot air onto a rotating drum, forming an entangled fibrous web on the surface of the drum. The process is shown schematically in Fig. 6.1.

Regardless of the production method and the nonwoven type, a concern regarding the use of nonwovens in lithium-ion battery separators is the relatively large pore size of these materials. For example, many

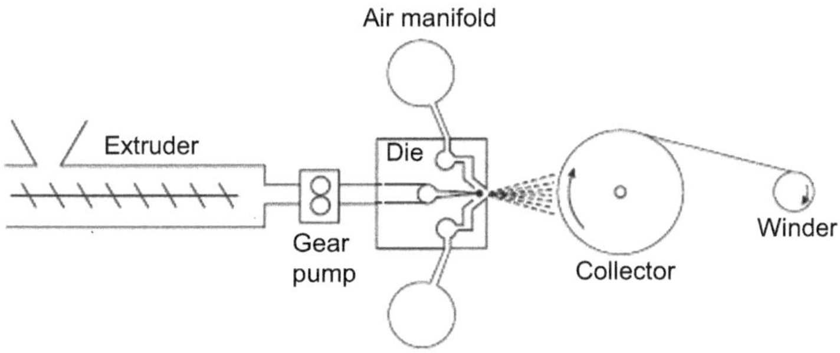

Figure 6.1 Diagram of melt-blown process.

nonwoven materials have pore sizes that are greater than 1 μm [1]. This is larger than the pore size usually specified for lithium-ion battery separators. The preferred average pore size for these separators is less than 0.1 μm [2]. That size is usually needed to prevent dendrites from penetrating the separators and producing holes. Thus, nonwoven separators may present safety issues when the battery is used and may not be able to prevent the occurrence of catastrophic thermal events in the battery.

Evonik and DeGussa addressed this concern by using a combination of ceramics and polymers, two materials normally completely incompatible because of their different temperature profiles [3]. That technology was acquired by Electrovaya and is not part of Litarion Evonik. The name of the separator is SEPARION; it is made from ceramics and a porous polyethylene terephthalate (PET) nonwoven that together form a homogeneous single-layer network. The ceramic coating is applied as a suspension onto the polymeric nonwoven support. Hence, the ceramic material is embedded into the PET nonwoven support material. There are two commercial SEPARION separators; they have a thickness of 21 μm (P20) and 28 μm (P30), respectively.

SEPARION claims to outperform trilayer and other ceramic-coated separators. It is designed for high-temperature stability that enables cells to improve safety performance [4]. Primarily because of its unique structure, the separator provides mechanical integrity at temperatures up to 240°C and a low shrinkage of <1% at 200°C, even when applied for several hours. SEPARION separators are designed to save time and cost in the manufacturing. Cell stacks containing the separator can be dried at high temperatures without melting or shrinking. The wettability features are meant to minimize the electrolyte filling time and pressure.

The high-temperature melt integrity of 240°C that is displayed by SEPARION separators is certainly a feature that makes them attractive for use in electric vehicle batteries. That melt integrity largely occurs because the separator is based on the use of PET combined with inorganic particles. PET has a melting temperature in the range of 260°C whereas the inorganic particles display outstanding high-temperature features. The measured shrinkage of <1% is another feature that contributes to the usefulness of SEPARION in electric vehicles. That feature means that it will function effectively at elevated temperatures to prevent short circuits. This is necessary for separators to provide safety to batteries used in electric vehicles.

One concern regarding SEPARION is that without adding another layer or more polymers, they will not provide shutdown to the battery. The batteries must rely solely on the high-temperature melt integrity provided by the separator for safety features. Thus, it is still unclear which safety mechanism, shutdown or high-temperature melt integrity, is more important to guarantee the safety of the battery.

The other feature of the SEPARION separator that is concerning is the relatively low strength it possesses. Owing to their open pore structure, nonwovens have relatively low mechanical properties, such as modulus and strength. Care must be exercised during handling and winding of the separator. This needs to be done to prevent the development of defects in the separator. Along with the low modulus and strength, the ceramic-modified PET separator is likely to have low puncture strength values. This may be a general concern with separators that contain a high loading level of ceramic particles. For example, membranes that contain a large amount of aluminum oxide have reported puncture strength values around 250–300 g. On the other hand, commercial separators from Celgard have puncture strength values around 350–450 g.

Because SEPARION separators involve a combination of dissimilar inorganic and organic materials, it is necessary to employ an adhesion promoter. Initial work employed a metal oxide sol in this regard [5]. The separators obtained using that type of adhesion promoter mostly have outstanding technical properties, but the metal oxide adhesion promoter of the ceramic particles limits flexibility of the layers to a specified maximum bending radius.

That issue was addressed in US Patent 9680141 [6] through the use of an adhesion promoter composed of a mixture of silanes and at least one thermally cross-linkable acrylic polymer. In addition, a carboxylic acid preparation and a defoamer component free of silicone oil are used to formulate the separator. The separator that is obtained has improved mechanical properties owing to the presence of the acrylic polymer in the adhesion promoter. The silane content of the adhesion promoter provides a significant increase over prior work in the ceramic content, and hence in thermal stability. The produced separators combine the advantages of purely organically bound separators with those in which purely inorganic binders are employed.

SEPARION was found to be a more versatile choice for lithium-ion batteries compared with nonfunctionalized polyolefins for several different hydrophobic ionic liquids as electrolytes [7]. Electrolytes with good solid electrolyte interface (SEI) building ability allow for stable performance with the SEPARION separator. This is because of the formation of an effective SEI as a combination of the initial SEI damage and the evolution in contact with the electrolyte material.

Dreamweaver developed another approach to incorporating nonwoven products into lithium-ion battery separators. They produced mixtures of nanofibers and microfibers that are intimately blended in high-shear mixing. The two materials are then cast together in a high-speed, high-quality, wet-laid nonwoven manufacturing process [8].

The microfibers in the mixture provide high strength and the ability to pack at high porosity. However, the pore size is too large to be useful in lithium-ion batteries. The nanofibers result in small pore sizes but pack at a high density level. This leaves little room for ions to pass through a membrane. In the separators themselves, the microfibers form a scaffolding over which the nanofibers drape, giving strength and dimensional stability as well as a small pore size. The nanofibers in the separator form a barrier to the growth of dendrites and other cross-contamination.

Two separators were produced using this general approach. The first is named Dreamweaver Gold. It provides safety to lithium-ion batteries. With a backbone of aramid fiber, the separator is stable up to 300°C and retains its shape and size even up to 500°C. It displays shrinkage values of less than 2% at 280°C when testing is done for 1 h. The melt integrity that is reported is greater than 200°C. Both of these properties make the separator a viable candidate for use in applications such as electric vehicles.

The other separator developed by Dreamweaver is called Dreamweaver Titanium. It is designed with low resistance in mind. The separator delivers the lowest internal resistance for high-rate cells. It has excellent wettability. The Dreamweaver Titanium separator comes in three different thicknesses: 15, 17, and 30 μm.

In general, the Dreamweaver separators provide high porosity in a uniform stable sheet. They have good chemical resistance. They possess a uniform pore structure with a narrow pore size distribution. The average pore size is larger than in other separators. Typical values range from 0.4 to 1.0 μm, and the specific value depends on which product is being discussed. For example, Dreamweaver Gold has an average pore size of 0.4 μm, whereas Dreamweaver Titanium products have average pore sizes that range from 0.6 to 0.9 μm, depending on which product is selected.

The Dreamweaver products display excellent wettability. Materials wet in just seconds, reducing processing time and increasing uniformity.

Thermal stability is above 200°C and the materials do not shrink or melt. This reduces the potential for shrinkage-induced short circuits. Finally, they display low electrical impedance and high porosity values. These effects lead to high rate capabilities.

It is generally believed that for lithium-ion batteries to continue to expand into the largest energy storage opportunities, the cost will have to come down. In that regard, Dreamweaver technology builds on decades of development. These are based on the establishment of high-speed, high-efficiency, and high-quality wet-laid nonwoven production processes. These processes are routinely used in high-quality filters and other applications. By drawing on this innovation and expertise, Dreamweaver separators can be made using existing equipment at rates that far exceed those used in other battery separator production processes. These production rates translate into lower production costs, which is a positive feature for reducing lithium-ion battery costs.

Both the Evonik and the Dreamweaver technologies can provide separators with stability at a higher temperature than is provided by polyolefin-based separators. For example, separators based on PP are limited in temperature performance by the melting of the polymer that occurs at 160−165°C. On the other hand, the Evonik separator is based on the use of PET, which melts at about 260°C, whereas the Dreamweaver separators use high-temperature polymers such as aramids. Use of these high-temperature polymers is not easily done with current dry process technology because such polymers do not allow for the development of the appropriate structure and morphology, for example. Thus, the technologies of Evonik and Dreamweaver allow for high-temperature polymers to be used, which is an advantage provided by those approaches. By using high-temperature polymers, the thermal stability of the separator can be significantly increased. This is an important feature for using these separators in electric vehicles.

On the other hand, the mechanical properties of separators based on using nonwovens will likely be lower than for other separators. This could potentially lead to handling concerns and issues regarding the separators. Care must be taken to avoid tears and breaks from occurring in the separator during fabrication. Modifications to the handling procedure may be necessary to achieve the most effective battery production. Such modifications are certainly possible, but recognizing this need is the first step in implementing these novel separators into lithium-ion batteries. Only through this recognition can batteries that contain these separators be optimized in terms of their performance. This is an example of the need to understand the battery structure and its fabrication completely. Furthermore, the effect of each component on processing and performance needs to be appreciated. The battery needs to be considered in its entirety, and not as single components. It also needs to be defined whether the electrolyte and electrode materials commonly used in lithium-ion batteries are the best choices for separators based on nonwoven materials.

Therefore, nonwoven materials in lithium-ion batteries offer the use of polymers other than polyolefins. High-temperature stability can be incorporated into the separator for new applications such as electric vehicles. However, because of the large pore size in nonwovens, it is necessary to develop effective ways to close or block some of the pores. Two ways to accomplish that effect have been described.

Although separators can provide potential advantages in terms of thermal stability, some of their unique features will require the battery production process to be modified to use these separators in most effectively. This is primarily because of the relatively low strength present in these separators. However, those issues can be overcome and the separators will likely find an increased number of applications in developments such as electric vehicles.

Electrospinning

Primarily because of unique properties such as the high porosity and high specific surface area, electrospun nanofibers have become the subject of many research projects for a wide range of applications. Among them are water, building, environmental, energy, automotive, and health uses. However, the relatively low production rate of a needle-type electrospinning process has been the main obstacle for transitioning the technology from the laboratory to industry. The mechanical complexity of the design and clogging of the needles during the process are two major obstacles to using a multineedle system for the industrial-scale production of nanofibers.

Electrospinning was developed to fabricate polyvinylidene (PVDF), PVDF-co-hexafluoropropylene, and polyacrylonitrile (PAN) membranes as separators for lithium-ion batteries [9]. The unique characteristics of the nanofiber web, such as a large surface area to volume ratio and high porosity, are different from what is usually obtained using either dry or wet process technology. Those features are crucial to increasing the ionic conductivity of membranes filled with liquid electrolyte. Because of these encouraging results, the production of nanofiber-based polymeric battery separators has potential and offers promise.

A brief description of the electrospinning process will be provided. In that process, superfine fibrous and porous membranes are produced by forcing a polymer solution through a spinneret with an electric field. Typically, an electric potential is applied between a droplet of a polymer solution, or melt, held at the end of a capillary tube and a grounded target. When the applied electric field overcomes the surface tension of the droplet, a charged jet of polymer solution is ejected. The discharged polymer solution jet undergoes an elongation process, which allows the jet to become long and thin. Meanwhile, the solvent evaporates, leaving behind a charged fiber. In the melt, the discharged jet solidifies when it travels in the air. The average diameter of the fibers produced by the electrospinning process is 50–500 nm.

DuPont introduced the first nanofiber-based polymeric battery separator that boosts the performance and safety of lithium-ion batteries [10]. Energain battery separators are produced from a polyimide into a web using a proprietary spinning process. Filaments are created with diameters of 200–1000 nm. The separators exhibit stability and low shrinkage at high temperatures. They are also highly saturable in liquid electrolytes. Use of the separators can increase battery power by 15%–30% and battery life by up to 20%, and can improve battery safety by providing thermal stability at elevated temperatures. With more battery power, it is possible to travel farther on a single charge and accelerate more quickly and safely. In addition, more battery power can reduce the number of batteries typically required in hybrid and electric vehicles.

One of the biggest issues with the Energain separators is the relatively high cost of the polyimide polymers that are employed. This high cost is largely associated with the monomers used to produce the polyimide. In general, the high price associated with high-temperature polymers is a concern with their use in battery separators and is an area that needs further attention.

Electrospinning technology has been demonstrated in battery separators for a number of polymers and combinations of different polymers. For example, polyimide (PI)-PET composite separators were produced by electrospinning [11]. The porosity, liquid electrolyte uptake, and room temperature ionic conductivity of the PI-PET composite membranes slightly decreased compared with a control PI membrane. On the other hand, the tensile strength of the PI-PET composite separator was higher than for the PI membrane. A decrease in the fiber diameter led to an increase in porosity as well as liquid electrolyte uptake of PI-PET composite membranes. The PI-PET composite membrane showed only about 2% thermal shrinkage at 180°C. Batteries that contained the PI=PET composite membranes had stable cycling performance as well as excellent rate capabilities.

Battery separators based on PAN-PI composites were prepared by electrospinning [12]. PAN membranes prepared by electrospinning had high porosity, electrolyte uptake, thermal stability, and lithium-ion conductivity. However, they also generally possessed poor mechanical properties. Results for the composite membranes showed that they possessed good thermal stability and exhibited better mechanical properties than pure PAN membrane. The addition of PI increased the porosity and liquid absorption rate. Furthermore, the composite membrane had high ionic conductivity, a wide electrochemical window, and excellent cycling performance.

Nanofibers fabricated by electrospinning have been used to construct battery separators with the structure PI-PVDF-PI [13]. These separators possess a thermal shutdown function using PVDF as the middle layer in the separator. On the other hand, the presence of PI as the top and bottom surface layers in the structure provide good thermal stability to the separator structure. The electrolyte uptake and ionic conductivity are excellent and result in superior performance in terms of impedance, discharge capacity, and cycle life. The self-shutdown function of the separator was observed within 10 min, which could serve as a safety mechanism to eliminate the thermal runaway issue in lithium-ion batteries. The effects of heating temperature and different times on the morphologies of each layer and the electrolyte uptake of the separator were characterized as well in that work.

In addition to that study, PVDF-PI membranes were prepared by a cross-electrospinning process [14]. That work employed two types of polymer solutions arrayed alternately in a straight line to form a composite membrane. The composite membrane showed positive results in terms of physical properties. In addition, cells assembled with the composite separators exhibited better electrochemical stability, lower interfacial resistance, and higher ionic conductivity compared with cells based on a commercial Celgard PP membrane. The results indicated that the bicomponent cross-electrospinning method can be extremely useful for designing future battery separators.

The electrospinning process was also applied to the production of battery separators from polymer other than polyimides. For example, electrospinning was used to prepare polytetrafluoroethylene (PTFE) nanofibers employed in a battery separator [15]. The results of that work showed that the PTFE nanofiber separator had a special fiber structure made from PTFE particles gathered along the fibers. Furthermore, the PTFE nanofiber separator had several advantages, including a suitable pore diameter, uniform pore size distribution, high porosity, and electrolyte uptake. The thermal stability of the PTFE nanofiber separator was much better than that of a conventional polyolefin separator. Cells equipped with the PTFE separator exhibited excellent rate performance as well as relatively stable cycle performance.

Another unique battery separator was prepared by combining an SiO_2 nanofiber membrane with a PVDF nanofiber membrane [16]. The mechanical strength of the composite nanofiber membrane was twice as high as the pure SiO_2 nanofiber membrane. In addition, at 200°C, there were almost no dimensional changes in the composite membranes. Furthermore, the composite separator exhibited fairly high ionic conductivity that significantly improved the performance of lithium-ion batteries. Batteries assembled with the composite membrane showed excellent capacity retention, which led to the conclusion that they were promising candidates for separators for use in high-performance rechargeable lithium-ion batteries.

In addition, electrospun polyvinyl alcohol (PVA)-SiO_2 separator membranes were prepared and evaluated for their performance in lithium-ion batteries [17]. PVA was used to prepare nanofiber-based membranes. A sol-gel technique was used to introduce SiO_2 into the PVA nanofibers. The PVA separator had excellent electrochemical properties with a highly porous structure. The use of SiO_2 further improved the porosity, electrolyte uptake, and ionic conductivity and reduced interfacial resistance. Lithium—lithium iron phosphate cells were assembled using PVA-SiO_2 separator membranes. They displayed high cycling and rate performance.

Development efforts to produce battery separators by electrospinning focused on two factors. First, high-porosity separators can be obtained by using electrospinning. This is a general necessity for separators that will be used in electric vehicle applications, because they will require increased power and energy. This is achieved in high-porosity values through the absorption of additional electrolyte material. In fact, as discussed, many separators based on electrospinning have increased electrolyte uptake features.

The second focus of much research on electrospinning of battery separators has been on developing materials that provide increased thermal stability over other separators. This is an important feature for the safety of lithium-ion batteries. This becomes a critical factor for larger batteries used in applications such as electric vehicles and energy grids. Separators produced by electrospinning often have higher thermal stability features that are achievable with polyolefin separators. This is possible because electrospinning can use high-temperature polymers such as polyimides in the separator production process. This is a positive feature of electrospinning in that a wide range of polymers can be used. This is in contrasted to both the dry and wet process technologies, for which a limited number of polymers are typically used. In addition, it is possible to incorporate inorganic materials such as SiO_2 into the formulations. The properties can be tailored and fine-tuned as needed.

Tailoring of separator properties is an important feature of electrospinning. This can be achieved by selecting the required polymer with the desired properties. Furthermore, by using appropriate processing parameters, it is possible to change the features of the separators as needed. A single formulation may be applied to multiple applications. This can have significant implications for the costs that are associated with implementing the process into commercial-scale operations.

Much more work needs to be done in that regard. There have been numerous interesting laboratory studies on the electrospinning of battery separators, but those encouraging initial results need to be expanded to larger-scale production levels. Important features to achieve that goal need to be defined. Then, plans should be developed to obtain those goals by producing an adequate amount of separator material for large-scale evaluations. This is the only way to determine any processing issues completely that may affect large-scale production of the separators. This is an essential feature and aspect of moving the technology from the laboratory to full-scale industrial implementation.

As part of that effort, a full investigation needs to be conducted of the economics of electrospinning. Operating speeds and efficiencies should be determined that allow for a cost-effective process to be used. Also, the cost of the polymer materials used in the process need to be considered. If high-temperature polymers such as polyimides are to be used, this must be factored into the overall process economics.

Separators made by electrospinning nanofibers can have high porosity and permeability as well as improved wettability by the electrolyte material. In addition, nanofiber layers with porosity values greater than 90% can reduce the resistance to ionic current and allow for the higher transport of lithium ions. As a result, a significant increase in battery power density, faster charging, and an increase in battery life can be provided by separators based on nanofibers. Electrospun nanofiber-based membranes are promising candidates for separators in lithium-ion batteries to enhance the transportation efficiency of lithium ions. Much promising research has been done on the topic; additional work should focus on developing those efforts into commercial products.

Beta-nucleated polypropylene

The first two sections of this chapter summarized the use of polymers other than polyolefins in battery separators. The technologies that were discussed for using those polymers to produce separators are much different from current dry and wet production methods. As such, implementing the technologies and using the corresponding polymers will be much different from current approaches being used.

On the other hand, the technology that will now be discussed involves using PP in the battery separator. The process employed to produce the separator involves the biaxial stretching of films produced using a special additive. As such, it draws on the well-known biaxial stretching process and techniques that are associated with it [18]. The general biaxial orientation process is used to produce PP films employed in food packaging applications as well as other uses.

PP can exist in several different crystalline forms. The semicrystalline structure is complex and depends to a large part on thermal history and processing conditions. The alpha phase is the most stable crystalline type and is the most common. The crystals are monoclinic in nature. The beta phase is metastable and the crystals are pseudohexagonal. It can be generated by adding specific nucleating agents. This form can be improved by crystallization at 130—132°C or by orientation with high shear. The presence of the beta phase generally increases ductility in the finished parts. Finally, there is the gamma phase, which is also metastable with triclinic crystals. This crystal form is not familiar, but it appears primarily in low—molecular weight PP. Important factors in its occurrence are high pressures and low cooling rates.

The transition from the beta phase to the alpha phase of PP is interesting. The beta phase occurs to a high level in the presence of a beta nucleant. In addition, crystallization under high shear conditions and in a thermal gradient are important. Finally, crystallization at of 90−130°C is important for the development of high levels of beta crystallinity.

The beta phase melts at about 150°C. Few beta nucleants are known. The beta phase transforms to the alpha phase upon stretching. It undergoes more uniform drawing than the alpha phase. It exhibits microvoiding during stretching, and the presence of those microvoids allows for the use of the films in battery separator applications.

The production of the separator is usually through a dry-stretch or oriented PP technique. In the process, the PP polymer is doped with the beta nucleator and the resin is then extruded. The precursor film is then formed and the precursor is stretched into the battery separator. To obtain the desired balance of physical properties, the precursor preferably has a beta crystal content of 45%−60%.

Several classes of materials are known to be specific beta-nucleating agents (eg, quinacridone pigments) [19]. In addition, a well-known class of beta-nucleating agents is Group I salts of dibasic organic acids. In addition, mixtures of calcium carbonate and a calcium salt of a dibasic organic acid were developed [20].

One of the first uses of the beta-nucleating agent approach to produce PP microporous membranes was in US Patent 5134174 [21]. The process described in that patent involves biaxial stretching of the starting film to create a microporous structure, and then heat-setting the resulting porous film under biaxial tension to achieve a further increase in porosity to impart thermal stability to the porous structure in the final film product. These final film products are produced from a uniform PP starting film with a high content of beta crystals obtained by extruding PP granules with nucleating agents. The patent also discusses the effect of different processing conditions on the development of the beta crystal phase in the film.

A conventional screw extruder equipped with a coat hanger die is deemed satisfactory as an apparatus for preparing the starting films. PP resin with the nucleating agent can be introduced into a hopper of the extruder. The resin is then melted and transferred by the screw to the die from which it is extruded through a slot die into a film. The film is then run onto a takeup or casting roll.

The temperature of the melt is no higher than about 100°C above the melting temperature of the PP. It should be no lower than the PP melting temperature. A preferred temperature range of about 180°C to about 250°C is quoted in the patent.

It is important for the cooling process of the slot-cast film to be controlled, to obtain uniform starting films with a higher content of beta crystals. The temperature that is appropriate for the cooling of the extruded film is about 80–135°C. More than one takeup or casting roll in various combinations can be used to improve the uniformity of the starting films. The temperature of the rolls needs to be well-controlled.

The uniform starting films are then subjected to biaxial stretching. In this process, the starting film is stretched in two directions perpendicular to each other. The draw ratios in the two perpendicular directions are preferably of equal value, or at least values of a small difference. The temperature for biaxial stretching is 80–140°C, preferably 90–130°C.

After the biaxial stretching process is completed, the microporous film is heat set at a substantially constant dimension. Heat setting is carried out at 90–155°C, preferably at 110–140°C. The period of heat setting treatment is preferably within about 0.5–5 min.

The films that are produced have a porosity greater than 20%, usually 30%–35%. The average pore size is 100–2000 Å, usually 200–800 Å. In addition, the mechanical properties of the microporous films are better than those of previous investigations. The tensile properties are uniform in all directions.

A deficiency of the films for use as battery separators was recognized in US Patent 6368742 [22]. For example, limitations of the films were disclosed to arise from low puncture strength values related to the stretching conditions as well as the pore size and porosity.

US Patent 6368742 [22] specifies that biaxial stretching should be in the machine direction (MD), transverse direction (TD), and optionally, a relax or stress relief step. The MD stretch temperature is 70–110°C, most preferably at 90°C. The MD stretch ratio ranges from 1.5 to 6.0; 4 is preferred. The TD stretching temperature is 110–140°C; 120°C is the most preferred temperature. The TD stretch ratio ranges from 1.5 to 6.0; ratios of 4–5 are most preferred.

Samples were made on pilot line equipment representing a conventional biaxially oriented PP production line. The line includes an extruder, a variable speed roll with temperature control that serves as the crystal formation section, and a draw frame consisting of both MD and TD stretch. The residence time, or the time on the casting roll, was about 30 s for all samples.

Samples that were produced generally had porosity values of 45%–50%. Most puncture strength values were greater than 400 g of force. Typical MD stretch ratios for the reported samples were 3–4 at temperatures of 120–125°C. The TD stretch ratios were 3.0–4.5 at temperatures of 120–140°C.

New Japan Chemical also obtained two patents for microporous film and the process for producing the film from beta-nucleated PP [23,24]. That work focused on some issues that may be encountered during the separator production process. For example, there is a strong tendency for pore formation to be impaired if the strain rate is high during the transverse stretching step in production. Sometimes, no pores form if stretching is carried out at a transverse stretching strain rate of at least 100%/s. However, using lower strain rates is not always desirable because it leads to lower productivity.

In addition to the problem of pore formation, another issue to be addressed with porous films is that breaks occur in the film during manufacturing. This problem appears to be less significant when beta PP crystals are used compared with a porous film containing a filler such as calcium carbonate. However, the results show that further improvement is still needed on this topic.

The two patents from New Japan Chemical state the need for a better understanding of the pore formation mechanism. Furthermore, the need to establish an industrially optimal manufacturing method suitable for this mechanism is discussed. Specifically, uniformity of the thickness of separators produced from beta-PP was inadequate. That led to unsatisfactory uniformity in the film's characteristics, such as permeability, tensile features, electrical resistance, and porosity. With that in mind, New Japan Chemical is attempting to provide a manufacturing process for the film that is resistant to breakage during manufacture. Two key features of the process are that the film can be manufactured in a stable manner and at a high strain rate.

Work by New Jersey Chemical uses an amide compound as a beta crystal nucleating agent. The extrusion temperature was found to be an important factor in having an acceptable thickness uniformity in the final film. In addition, the neck-in ratio during longitudinal stretching was adjusted to specific levels. Both of these process adjustments had the effect of making the film less prone to breakage during the transverse stretching step and allowed an increase in the transverse stretching strain rate.

In addition to those process improvement steps, an annealing treatment was added under specific conditions between longitudinal stretching steps.

In this case, pore formation is further promoted and the properties of the resulting biaxially stretched film are further improved. The strain rate during the transverse stretching step can be even higher without affecting the breaking resistance of the film.

It is believed that the annealing treatment allows the beta crystals that remain after longitudinal stretching to undergo a transition to alpha crystals. If the annealing temperature is lower than 130°C, the transition from beta crystals to alpha crystals is inadequate. On the other hand, it is less preferable to perform annealing at a temperature higher than 160°C because the alpha crystals will melt and the orientation will be disturbed. It is undesirable for the annealing time to be less than 1 s because the crystal transition from beta crystals to alpha crystals is incomplete. Crystal formation is almost complete when the annealing time is approximately 300 s.

Using the methodology of the patents, it is possible to increase the transverse stretching strain rate. This is important because the transverse stretching strain rate can greatly affect pore formation and film breakage. For example, in a conventional manufacturing approach, if the strain rate is increased to 100%/s, there is a marked drop in air permeability. In addition, the likelihood of breakage increases. A strain rate of 17%/s or less is usually recommended. However, using the technology of the patents, it is possible to produce a porous film with good air permeability and little breakage even using high strain rates of 100%–300%/s.

The separators that are produced have good thickness uniformity. Thus, there is no substantial variance in separator properties from place to place in the film. The good thickness uniformity is also advantageous in terms of production stability.

There have also been academic investigations into the production of pores in separators that are based on using beta-nucleated PP. For example, the pore formation mechanism of beta PP films during stretching was investigated [25]. In situ small angle X-ray and wide-angle X-ray diffraction measurements were used to define the relationship between the initial microvoids and the evolution of lamellar structures in the initial cast films. Initially, cavities or microvoids originated from the lamellar separation at the weak boundary between the beta lamellae and the adjacent amorphous phase. Lamellae on beta transcrystals preferred to be separated along the MD. On the other hand, twisted and inter-cross-linked beta lamellae were more likely to separate along the TD.

Stretching in both directions affected the second stage of pore formation, and the growth of microvoids could be promoted along the TD. For the stretching of samples along the TD, microporous membranes with higher porosity levels and more homogeneous microporous membranes were obtained by subsequent biaxial stretching. Therefore, the stretching methods were optimized to prepare microporous membranes with excellent performance features.

Another investigation into pore formation during the biaxial stretching of beta PP was performed by Ding et al. [26]. That work researched the pore formation mechanism of different kinds of beta PP spherulites. There does not appear to be a relation between deformation unity and pore uniformity in the beta PP. The transverse stretching process was examined in great detail and pore formation and evolution were studied. The key result of the studies was that pore size distribution depends on the size and arrangement of the beta lamellae.

In another study [27], the pore formation behavior and pore performance of beta-nucleated PP microporous membranes were studied using both differential scanning calorimetry and wide-angle X-ray diffraction. Samples that contained different levels of a PP random copolymer were examined. The presence of the random copolymer slightly inhibited the formation of the beta crystals and significantly reduced the melting point. As the content increased, the mechanical performance of the samples increased, but the deformation uniformity of the sample decreased. In particular, the deformation of samples with a high copolymer content was more inhomogeneous and dense regions were generated during stretching. This had an overall effect of broadening the pore size distribution of the sample and reducing interconnectivity between micropores. Samples with a low random copolymer content and high copolymer content have different amounts of sample porosity and pore connectivity at elevated temperatures. The different membranes will show completely different behavior at high temperatures.

There has also been commercial activity in the production of battery separators based on beta-nucleated PP. Treofan produced a product named Treopore, based on beta-nucleated PP technology [28]. The separator is described as possessing significantly higher thermal stability compared with other polyolefin separators. With those improved properties, the separator film will be particularly useful in applications such as electric vehicles and electric mobility in general.

A primary advantage provided by beta-nucleated PP technology for battery separators is that it draws on the well-known biaxial stretching process of PP. The equipment and associated technologies are well-developed.

As such, much knowledge and expertise that can be drawn upon to implement the technology.

Biaxial stretching of film is known to be a low-cost, inexpensive production process. This is because of the wide width of the final film that is produced. Thus, it is possible to produce a large quantity of film from a relatively low quantity of starting polymer material.

Nevertheless, most commercial biaxial orientation processes involve orientation levels of five times in the MD and up to 10 times in the TD [18]. There is an increase in the area dimension of the film by a factor of approximately 50. On the other hand, the stretching ratios used to stretch the products from beta-nucleated PP use area dimensional changes on the order of 20. Therefore, less film is made for each unit of polymer starting material. The production costs for separators based on beta-nucleated PP will be higher than for standard biaxially oriented PP films. Thus, there will not as large a positive impact on the overall economics of the process.

Also, typical biaxial orientation processes can operate at line speeds approaching 1000 ft/min [18]. A huge amount of film can be produced in a short time. On the other hand, the production process for beta-nucleated based separators needs to run slowly because it is necessary to have a certain residence time on the film casting roll to develop the beta crystals fully. This residence time limits the speed at which the stretching line can be operated. This limitation provides a maximum to the operating line speed, and hence to the film output rate. Thus, the economics of the process will be negatively affected.

Despite possible economic advantages of using biaxial stretching of membranes based on beta-nucleated PP, it is apparent that production costs will not be as low as for standard biaxially oriented PP products. It is necessary to perform an economic analysis of the production process for actual parameters in terms of both the final film width and the line speed that can be achieved. Only through such an analysis can the true economic advantages of the biaxial orientation process be obtained.

A potential issue with using the beta-nucleated PP is the availability of the beta-nucleating agent in adequate commercial quantities. Mayzo [29] addressed that concern by developing masterbatches containing the beta-nucleating agent. The masterbatch can be added at the extruder hopper to a nonnucleated resin to achieve the benefits of beta-nucleation. Mayzo provides four masterbatches containing a beta-nucleating agent. The various masterbatches differ in the features of the carrier resin that is used.

The nucleating agent that is employed in the masterbatches is described as a proprietary beta-nucleant formation.

No complete study has been published showing which beta-nucleating agent is most effective for producing battery separators. It may be that the optimum material is not the same one that is the best nucleating agent for products being developed for injection molding applications. More work with the biaxial stretching process needs to be done to guarantee that the optimum formulation is being used.

Some work has been performed on defining the best processing parameters for the biaxial stretching of beta-nucleated PP-based films. Certainly, the optimum processing features will be different for various beta-nucleating agents. Additional work related to a correlation between the beta-nucleating agent and the structure of both the cast film and the final film-oriented film product needs to be conducted to determine the best overall processing conditions for each beta-nucleant. Such an analysis will allow for the optimum use of the beta-nucleated PP technology in battery separators. Also, it will permit the efficient scale-up of the process to production levels necessary for the use of the separators in electric vehicles.

Along those lines, in 2017, Treofan signed a contract with a Chinese manufacturer of lithium-ion batteries for use in electrically powered buses [30]. The goal of the contract was to reach 28 million m^2 of separator annually in 2020. That amount of film is roughly equivalent to the area of separator film needed for 50,000 standard battery packs, each containing 56 kWh of energy. The contract marked the first large-scale entry of a European separator manufacturer into this market.

One final aspect of the separators based on beta-nucleated PP is the ability to provide shutdown features to the battery. In the monolayer PP film, shutdown will not occur until about 160°C, the melting point of the PP. To provide additional safety to the batteries, lower temperature shutdown in the separator is required. This can be achieved by incorporating a separate layer into the separator for the shutdown function. However, this will make the separator structure more complex and will likely increase production costs. On the other hand, the use of a shutdown feature would provide added safety, and that feature may justify the additional cost to the separator.

Conclusions

The development of electric vehicles has been the main force for establishing new technologies to produce of battery separators for lithium-ion batteries. The large size of the batteries means that the safety of the battery is more crucial than in usual consumer electronics applications. The separator can affect the safety of the batteries through its ability to keep electrodes separated at elevated temperatures. This is achieved by the separator having high-temperature thermal stability features.

Current dry and wet process technologies are limited in terms of the types of polymers they use. Polyolefins, both PE and PP, are used in those processes. Because of the use of those polymer classes, thermal stability, and hence the safety that can be achieved, are limited.

A key advantage of the first two technologies discussed in this chapter is that they offer the opportunity to use polymers other than polyolefins. High-temperature stability can be imparted to the separator by selecting the appropriate polymer or combination of polymers. This can be achieved without modifying the polymers used. For example, polymers such as PET and aramids can be used in separators made using the technologies.

Nevertheless, to use nonwovens in battery separators for lithium-ion batteries, it is necessary to adjust the average pore size. This is because the pore size in the initial nonwoven material is generally larger than what is desired. As discussed in this chapter, this modification of pore size can be implemented by using either an inorganic filler material or another polymer structure. Both approaches are effective ways to reduce the pore size so that the base nonwoven product can be used in the lithium-ion battery.

On the other hand, the other technology discussed in this chapter, beta-nucleated PP, addresses the production costs of the separator. It is a cost-effective means to produce battery separators. However, the economics associated with the actual production process need to be better defined to define the advantages that are provided. Such an analysis will provide important insight into the definite economic advantages provided by the approach.

Of course, each discussed technology has disadvantages as well. Among the most significant is the need to modify the battery fabrication process to deal with separators that have handling features different from those of standard separators. The key point on that subject is the realization of the overall need to adjust the process and then make the necessary changes. That strategy will allow for the most effective use of these separator materials. The realization should be regarded as an opportunity to extend the current battery fabrication process to deal with novel separators that are produced. However, this will require an additional effort on the part of battery manufacturers to implement these new separator structures satisfactorily into the lithium-ion batteries that use them.

New applications will continue to drive the need to develop novel battery separator structures. A single separator will likely not be optimal for all emerging uses. The acceptance of that fact will be important in future work on novel battery separators for use in lithium-ion batteries.

References

[1] W. Yi, D. Xiaoli, Y. Yunzhen, L. Yun, H. Jian, Z. Huanyu, Textil. Res. J. 82 (16) (2012) 1659.
[2] G. Venugopal, J. Moore, J. Howard, S. Pendalvar, J. Power Sources 77 (1) (1999) 34.
[3] Altech Chemical Limited, High Purity Alumina (HPA) Application on Lithium Ion Battery Separator, May 11, 2020.
[4] C. Morris, Newswire, The Tech, December 18, 2015.
[5] EP 147845B1, Assigned to Evonik Operations GmbH, May 5, 2005.
[6] U.S. Patent 9,680,141, Assigned to Litarion GmbH, June 13, 2017.
[7] M. Kirchofer, J. Von Zamory, E. Pailard, S. Passerini, Int. J. Mol. Sci. 15 (8) (2014) 14868.
[8] dreamweaverintl.com/technology.html.
[9] Y. Li, Q. Li, Z. Tan, J. Power Sources 443 (2019) 227262.
[10] www.greencarcongress.com/2010/08/dupont - introduces new separators for Li - ion batteries.
[11] J. Ding, Y. Kong, P. Li, J.R. Yang, J. Electrochem. Soc. 159 (2012) A1474.
[12] L. Li, P. Liu, Q. Fu, Y. Gong, J. Mater. Res. 34 (4) (2019) 642.
[13] D. Wu, et al., Electrochim. Acta 176 (2015) 727.
[14] W. Chen, Y. Liu, Y. Ma, J. Liu, X. Liu, Mater. Lett. 133 (2014) 67.
[15] Li, et al., J. Appl. Polym. Sci. 135 (29) (2018) 46508.
[16] Y. Xu, et al., J. Nanomater. (2020), 3879040.
[17] M. Yanilmaz, J. Text. Inst. 111 (2020).
[18] T. Mark, DeMeuse, Biaxial Stretching of Films, Woodhead Publishing, 2011.
[19] EP 0177961A2, Assigned to PCD Polymers GmbH, 1988.
[20] US 201000010168A1, Assigned to Borealis Technology, 2010.
[21] U.S. Patent 5,134,174, Assigned to Institute of Chemistry Academia, 1992.
[22] U.S. Patent 6,368,742, Assigned to Celgard, Inc., April 9, 2002.
[23] U.S. Patent 7,235,203, Assigned to New Japan Chemical, June 26, 2007.
[24] U.S. Patent 7,682,689, Assigned to New Japan Chemical, March 23, 2010.
[25] Wu, et al., Polymer 163 (2019) 86.
[26] L. Ding, R. Xu, L. Pu, F. Yang, T. Wu, M. Xiang, Mater. Des. 179 (5) (2019) 107880.
[27] G. Xu, L. Ding, T. Wu, M. Xiang, F. Yang, J. Polym. Res. 26 (2019) 27.
[28] www.treofan.com.
[29] P. Jacoby, Plastics Addit. Compd. 9 (3) (2007) 32.
[30] www.plasteurope.com/news/TREOFAN- t236566/.

Further reading

[1] R. Waterhouse, J. Emanuel, J. Frenzel, D. Lee, S. Peddin. Y. Patil, G. Fraser - Bell and R.W. Pekala, The 29th International Battery Seminar Exhibit, March 12 - 15, 2012.
[2] C. T. Love, J. Power Sources 196 (2011) 2905.

CHAPTER 7
Ceramic-coated separators

Introduction

Coating conventional polyolefin separators with ceramic layers is considered to be one of the most effective and economic ways to enhance the safety of lithium-ion batteries. This is because ceramic coatings can suppress shrinkage and deformation of the polymer separator at elevated temperatures. The separator is able to function effectively to keep the electrodes apart and prevent short circuits that can lead to catastrophic events such as fires and explosions.

The use of ceramic coatings on the base separator increases the thermal stability of the separator. An increase in thermal stability in this context means that thermal shrinkage of the separator decreases at elevated temperatures compared with uncoated separators. This is the primary reason why electric short circuits occur, and the possibility of fires occurring is reduced. These effects are especially significant for large batteries used in electric vehicle applications. There is a larger potential for thermal runaway effects in these batteries than there is in smaller batteries used in consumer electronics.

Thus, the presence of ceramic coatings can extend the temperature performance of the separator. This is true of both polyethylene (PE) and polypropylene (PP) separators made using both the dry and wet process technologies. Ceramic coating technology has been applied to both of these types of polyolefins. A discussion of all of these efforts is provided during the remainder of this chapter.

The presence of a ceramic coating can also significantly affect both the physical and chemical properties of battery separators. In addition to the safety mechanism just mentioned, the coating layer may reduce or prevent dendrite punctures as well as the lithium plating. Furthermore, ceramic-coated battery separators offer the advantage of high electrolyte absorption.

This chapter summarizes developments in ceramic-coated separators. The primary focus of the chapter is on the structure of the separator and how use of the separator affects overall battery performance. This is a key feature of the use of ceramic-coated separators that needs to be quantified in

Polymer-Based Separators for Lithium-Ion Batteries
ISBN 978-0-12-820120-6
https://doi.org/10.1016/B978-0-12-820120-6.00001-5

detail. There will also be a discussion on the different types of inorganic fillers that can be used in the ceramic coating layer and how the different structures of inorganic particles can affect the performance of the batteries.

Ceramic coating basics

The formulation of a ceramic coating typically contains at least three components: inorganic particles, a solvent, and a binder polymer that helps with adhesion of the inorganic particles to the base polyolefin substrate. The presence of the binder polymer is required because the inorganic particles and polyolefin are chemically dissimilar. Different binder polymers have been used in different studies. The effect of the different binders will be discussed later in this chapter.

A common method for applying the ceramic coating to the separator is dip coating, in which there are generally five steps. The first involves immersing the base substrate into the solution of the coating material at a constant speed. Then, the substrate remains inside the solution for a period of time and then is pulled up. The thin layer is deposited on the substrate while it is pulled up. The withdrawing step is carried out at a constant speed. The speed determines the thickness of the coating, in which faster speeds give thicker coatings. At this point, excess liquid is drained from the substrate. Finally, the solvent evaporates, forming the thin coating layer. When volatile solvents such as alcohols are used, evaporation starts during the deposition and drainage steps in the coating process. These steps are carried out one directly after another.

Numerous features contribute to determining the final state of the dip coating: the functionality of the initial substrate surface, submersion time, withdrawal speed, number of dipping cycles, solution composition, concentration and temperature, number of solutions in each dipping sequence, and the environmental humidity. The dip coating technique can give uniform, high-quality film products.

Dip coating of battery separators can be done with materials other than ceramics [1]. For example, work by Song et al. [1] involves studies of coating PE separators with a substance from the polyimide family using a simple dipping method. The PE separators are provided with improved mechanical strength and increased resistance against thermal shrinkage at elevated temperatures. However, batteries that contain the polyimide-coated separators exhibited electrochemical properties almost identical to those of cells with the base PE separators.

In addition to dip coating, coatings can be applied using other technologies. These include coating knives, gravure, curtain, and spray approaches. However, dip coating appears to be the most widely used coating technology. One technology that has shown promise to replace the dip coating process is microgravure coating. In that process, a small-diameter gravure roll equipped with patterns is used [2]. The roll is partially submerged into a pan of solution. The roll rotates in the direction opposite the film web and the cells absorb the coating solution. The flexible doctor blade lightly smoothes off the excess and the measured solution in the cells is then transferred to the web.

There are several advantages to this coating technology. First, it provides exceptional uniformity with 1%—2% tolerance levels. It can be used to coat low-viscosity solutions that are both aqueous and solvent-based. The technology is suitable for evenly coating suspensions with rigid particles.

The actual formulation used in ceramic coating depends on which inorganic filler is used in the coating. That decision will dictate which solvent is needed in the formulation. The inorganic filler needs to be soluble in whatever solvent is chosen. In addition, selection of the appropriate binder material depends on which inorganic filler is being used. Certain binder materials will be more effective than others for use with certain inorganic materials. For example, the detailed effects of high—melting point binders on the thermal behavior of composited separators as well as the electrochemical performance was studied [3]. In that work, a case study using a poly(vinyl alcohol) (PVA) binder for the ceramic coating with aluminum oxide for polyolefin separators was done. The effects of using PVA and a commonly used low—melting point poly(vinylidene fluoride) (PVDF) were compared. The use of PVA was more effective in improving the thermal stability of the separators than was the PVDF. Increasing the PVA content increased the pore tortuosity and decreased the wettability of the coating layer, compromising cell performance. The results of the study suggest that to obtain high-performance composite separators, it is necessary to use a binder with a high melting point, good electrolyte wettability, and strong adhesion with the coating particles. Furthermore, the binder content should be well-controlled. The construction of a coating layer with low tortuosity will be beneficial to battery performance. The study provides valuable insights into the design of high-performance composite separators with improved safety features.

A study was performed to determine how different inorganic materials affect the properties of the separators in terms of both thermal stability and battery performance [4]. The study involved coating a commercial PP separator with four inorganic materials: include Al_2O_3, SiO_2, ZrO_2, and zeolite. All inorganic coatings provided improved thermal stability to the separator at different levels. The coating layers added 28%–45% to electrical resistance compared with the pure PP separator. All of the batteries prepared with the coated PP separators had the same electrochemical performance as cells that contained the uncoated separator. This was true of both the rate capability and capacities at different temperatures.

One of the most common ceramic materials used in ceramic coating layers is aluminum oxide [5–7]. In Shi et al. [5], a ceramic layer consisting of Al_2O_3 powder, carboxymethylcellulose, and a styrene–butadiene rubber mixed binder was applied onto one side of a PE separator. During preparation of the coating, water was used as the solvent and the binder mixture was applied to obtain better thermal stability. The results of the study showed that the ceramic-coated separator with the mixed binder had excellent thermal stability, good wettability, and high uptake of the electrolyte solution. Cells with the ceramic-coated separator showed excellent stable cycle performance.

Work by Liu et al. [6] produced separators containing Al_2O_3/SiO_2 ceramic layers. SiO_2 was used to form the porous network and Al_2O_3 was evenly distributed in it. The composite separator had lower thermal shrinkage and higher mechanical properties. Because the composite layers can retain additional liquid electrolyte, the composite separator had higher electrolyte uptake and ionic conductivity. Battery performance, including rate capability and cycle performance, was also investigated.

An et al. [7] used both an Al_2O_3 nanopowder layer and an electrospun PVDF nanofiber layer on the two sides of a PE separator. The presence of the Al_2O_3 layer provided excellent thermal stability, as shown by low shrinkage at 180°C and the absence of meltdown up to 200°C. The presence of the Al_2O_3 seemed to cause a loss of performance, but that was overcompensated for through the use of the PVDF nanofiber layer.

Shi et al. [8,9] coated aluminum oxide onto a PE separator using polyimide as the binder polymer in the formulation. The ceramic coating improved the thermal resistance and electrochemical characteristics of the PE separator by developing a microstructure with interconnected submicron-size pores on its surface. The hydrophilic nature of the polyimide and the alumina nanoparticles enhanced electrolyte uptake and ion conductivity, which in turn increased capacity retention.

Silicon dioxide or silica particles were also used to modify battery separators [10–13]. Yang et al. [10] coated core-shell structured silica-poly(methyl methacrylate) (PMMA) submicrospheres on one side of a PE separator. The functional ceramic-coated separator possessed both thermal stability and high electrolyte stability. The SiO_2 core particles protected the separator from thermal shrinkage. In addition, the presence of the PMMA shells functioned to stabilize the electrolyte. As a result, the separator showed considerable wettability of the liquid electrolyte and outstanding electrolyte retention at elevated temperatures. Moreover, the coated separator improved the safety performance of cells by preventing them from experiencing internal short circuits at high temperatures.

Fu et al. [11] studied the formation and deposition of SiO_2 compared with deposition using SiO_2 powder suspensions in solution to form a ceramic coating layer on PP separators. For both approaches, PVDF-hexafluoropropylene (HFP) was used as a binder and acetone was used as the solvent. The coated separators showed significant reductions in thermal shrinkage and improvements in tensile strength, contact angles, electrolyte uptake, and battery performance compared with the uncoated PP separators.

Prasanna et al. [12] tested stretched separators coated with SiO_2 ceramic layers. Coin cells were prepared by sandwiching the separators between the anode and cathode. The AC impedance and cycling performance of the cells were evaluated. The coated separators were observed to have superior ionic conductivity and tensile properties. Cells prepared with the coated separator had slightly higher discharge capacity and a better capacity retention ratio than did cells with uncoated separators.

Titania or titanium dioxide has also been used in the ceramic coating of lithium-ion battery separators [13]. Both improved thermal and dimensional stability are achieved by depositing the TiO_2 in the battery separators. The observed improvement can be attributed to the formation of a robust skeleton that stabilizes the separators. This imparts a superior insulation and mass transport barrier against volatile compounds formed during thermal decomposition. The TiO_2-coated separator features a well-developed porous structure that allows for favorable liquid wettability and a high mass uptake of electrolyte. The amount of TiO_2 that is deposited has a crucial role in facilitating the battery performance. Factors that are affected include high ionic conductivity, low inner resistance, high operating temperature, high energy density, and excellent cycling behavior.

Nickel oxide has also been used to coat separators [14]. PE separators uniaxially stretched 300% were dip-coated with sol-gel synthesized nickel oxide particles. The dip-coated stretched PE separator has high ionic conductivity and a relatively low MacMullin number. The coated separator has lower thermal shrinkage behavior even when it is exposed to high temperatures. These features enhance the safety of lithium-ion batteries.

In addition to the use of these common inorganic fillers, hydroxides of magnesium and aluminum have also been used [15]. The composite separators were prepared by adding ceramic coating layers that composed of one of the chosen metal hydroxides and a PVDF-HFP binder to PE separators. Both of the metal hydroxide-composite separators showed fire-retardant properties that helped to suppress thermal dimensional changes in the PE separators occurring at high temperatures. Primarily because of the hydrophilic nature of metal hydroxides, cells that used the composite separators achieved significant improvements in rate capability. Those improvements were closely associated with improved wettability for liquid electrolytes. The electrochemical stability of the magnesium hydroxide composite separator was comparable to that of the PE separator. However, the aluminum hydroxide composite separators were electrochemically unstable. This points to the significance of the high levels of capacity retention of cells that use magnesium hydroxide-based composite separators.

Ceramic-coated separators have enhanced properties compared with the polyolefin separator, but because of the low affinity between the particles and the polyolefin substrate, the coating adhesion strength can be low. This is issue is partially addressed by using a binder polymer. Another approach is to use plasma radiation to enhance the surface hydrophilicity of the substrate and nanoparticles. Plasma radiation can easily introduce electrolyte affinity and polar groups onto the surface of a hydrophobic polyolefin separator. Among the chemical groups that can be introduced are OH, −COOH, R−O-R′, and-NH_2. The actual groups that are introduced depend on the type of reactive gas used in plasma treatment. One advantage plasma treatment has is that it is more easily applied compared with other methods. For example, Jeon et al. [16] used plasma treatment to attach alumina particles onto the surface of a PE separator. After plasma treatment, the surface of the PE separators changed from hydrophobic to hydrophilic and the pore structures on the separators widened. The polar functional groups derived from the plasma treatment interacted with the hydroxyl groups of water-soluble polymeric binders in the aluminum oxide coating layers. This improved the adhesion strength between the PE substrate and the ceramic coating layer. As a result of this effect, cells based on plasma-treated ceramic-coated separators exhibit superior power capability and cycle performance.

Along with the plasma technique, grafting of nanoparticles onto the surface of the separator substrate by electron beam radiation has been adopted by researchers primarily because of its simplicity and easy modification. For example, X. Zhu et al. [17] grafted nanotitania particles onto the surface of a PE separator by electron beam radiation. The TiO_2-grafted PE separator had a thickness and pore structure similar to those of the PE separator. In addition, the grafted separator exhibited strong dimensional thermal stability and better electrochemical performance compared with the bare PE separator. This implies that the separator is beneficial for constructing safer lithium-ion batteries.

Furthermore, X. Jiang et al. [18] grafted a vinylsilane coupling reagent onto the surface of a PE separator through electron beam irradiation. The coupling agent then underwent a subsequent hydrolysis reaction and an ultrathin Al_2O_3 layer was grafted onto the surface of the PE membrane. The ceramic-grafted separator showed almost no shrinkage at 150°C. In addition, the contact angle to the electrolyte decreased compared with the bare PE separator. Finally, cells that used the grafter separator had better cycling performance and rate capability.

Choi et al. [19] grafted silica nanoparticles onto the surface of a PE separator. The silica nanoparticles formed a strong covalent bond with the hydrophobic polyolefin separator. This effect increased the uptake of the electrolyte and ionic conductivity. Also, owing to the covalent bond that formed, the thermal stability of the separator increased. These effects are largely the result of the newly formed Si—O—C or Si—CH_3 bonds that enhance electrolyte/separator compatibility. This may provide a better environment for ionic transport between electrodes, which can lead to better charging and discharging behaviors in batteries that contain the separators.

In addition, Liu et al. grafted polyacrylamide onto the surface of a PP separator and then coated silica particles by the sol-gel method [20]. Compared with the unmodified separators, the modified separators had better thermal stability and superior wettability. Shrinkage of the bare PP and the grafted separators weas 70% and 12% at 150°C, respectively. The modified separator improved thermal stability and the wettability and the thickness of the separator increased slightly. Also, cells with the grafted PP separator had better electrochemical performance than did those with the bare PP.

In addition to polyolefin substrates, ceramic particles were coated onto separators that were electrospun from other polymers. The electrospinning technology was described in Chapter 6. This approach could produce highly hydrophilic separators and thus enhance the features of lithium-ion batteries.

Zheng et al. [21] produced hybrid membranes with highly loaded silica nanoparticles. Good strength and flexibility of the hybrid silica membranes are provided by interpenetrated polymer nanofibers as a mechanical skeleton. The hybrid silica membranes have excellent dimensional stability at 200°C and high electrolyte uptake. Coin cells made with the hybrid membranes have superior discharge capacity and cycling performance at high temperatures up to 120°C.

In other work, Wu et al. [22] fabricated composite membranes made from ultrafine PVDF fibers that were coated with Al_2O_3 nanoparticles by a dip-coating process. The separator showed less than 2% shrinkage when tested at 140°C for 1 h. The fabricated membrane had a measured porosity of 55.8%. The alumina nanoparticles covered the fibrous separator with the use of polyethylene oxide as a binder material. The electrolyte uptake was reduced primarily owing to the reduction in pore size compared with the uncoated PVDF separator, but it was superior to that of a PE membrane.

Finally, Li et al. [23] prepared a composite separator by coating cubic Al_2O_3 nanoparticles on a nonwoven PET by a simple dip-coating process. The composite separator that was produced had excellent thermal stability. Compared with a commercial Celgard 2500 separator, the fabricated composite separator had higher porosity, electrolyte wettability, and higher ionic conductivity. In addition, batteries made using the separators showed a slight improvement in capacity compared with cells made with a PE separator.

The discussion provided in this section of the chapter has shown that there have been numerous fundamental studies on the ceramic coating of battery separators. The overall motivation for this work is to enhance both the electrochemical and physical characteristics of lithium-ion batteries. The most commonly used inorganic particles in the ceramic coating layer are alumina, titania, and silica. These materials have the ability to be dispersed in a solvent, possess a high surface area, have a hydrophilic nature, and can provide the separator with high thermal stability features.

Their use generally provides the separator membrane with mechanical strength, improves electrolyte uptake, provides electrochemical stability, and enhances the cycling capacity of the batteries that contain them.

Coating methods that have been used include dip coating, the sol-gel method, plasma treatment, and grafting. Substrates other than polyolefins have been used in the coating studies. The dip-coating method is common and more practical because of its easy method of application. On the other hand, it involves the use of organic solvents, which adds to the production cost of the separators. Also, the sol-gel method is an approach that can be used to reduce the cost of the coating by eliminating the solvent.

To deal with the hydrophilic nature of the polyolefin substrate, polydopamine can be applied on the surface before or after inorganic particle coating to enhance the hydrophilic character of both the separator substrate and the coating. Furthermore, plasma radiation can be used to graft inorganic particles onto the surface to enhance the strength of the coating and attain uniformity during the coating process. On occasion, the particles are modified to enhance their hydrophilic character or cross-linking before the coating is applied.

The inorganic coating can also be applied to electrospun separators. The purpose of this approach is to have a coating substrate with a reduced cost. The resultant separators have high porosity, mechanical strength, and electrochemical stability.

The coating approach increases interfacial resistance because of an increase in separator thickness. Also, there are chances of the nanoparticles detaching during fabrication of the lithium-ion batteries because the binder swells in the electrolyte. This drawback could be minimized by enhancing the affinity of the polymer and particles for each other and for the electrolyte. These effects can provide better capacity in the batteries.

A primary advantage of the ceramic coating approach is an improvement in the thermal stability of the separators. This effect has often been reported, and many of the articles quoted in this section of the chapter cite this situation; it seems to be a general effect. The magnitude of the effect depends on the actual formulation used to produce the ceramic coating. Also, the method of applying the coating seems to affect the properties of the separator.

One aspect of the improvement in thermal stability that is observed with the separators that is not often discussed is exactly how that improvement affects the performance of lithium-ion batteries. Improvement in the thermal stability of the separator means that the separator has low shrinkage at elevated temperatures. The electrodes will remain separated at these elevated temperatures and prevent internal short circuits. However, no studies are available that define quantifiable enhancements in battery safety that are possible by achieving an improvement in a certain level of thermal stability. For example, it is unclear at what temperature minimal shrinkage is required to guarantee battery safety. Ceramic-coated separators have been developed that display different levels of shrinkage at elevated temperatures, but it is not obvious how those different shrinkage levels actually translate into battery safety levels. In addition, different battery safety tests and evaluations might show different correlations with thermal stability results. It is necessary to establish and define those correlations to fully understand the advantages provided by ceramic coating technology. Results that are clearly quantified in that regard need to be summarized and recorded.

The entire area of structure—property—processing relations for ceramic-coated separators is a topic that requires additional study. For example, it has not been established whether there is an optimal inorganic filler for use in the ceramic coating layer to provide the highest level of battery safety. It may be possible that the mere presence of the inorganic filler in the ceramic coating layer improves the thermal stability of the separator.

Furthermore, within a particular type of inorganic filler, such as aluminum oxide, it needs to be determined whether there is an optimal particle size or particle size distribution to be used. Certainly, large particles are undesirable because of the negative impact they can have on the battery fabrication process using the separator. However, it is unknown whether a nanoparticle size is preferable to use in the ceramic coating layer. Also, it needs to be determined how different particle sizes affect the overall battery safety features.

For a specific inorganic filler, the best binder polymer needs to be defined. Much previous work involved the use of a standard PVDF binder in the ceramic coating layer. However, it was also shown that high-temperature polymers such as PVA can provide advantages to the ceramic coating layer. Other high-temperature polymers, such as polyimides, have also been used as the binder polymer in the ceramic coating layer. Any advantages provided by using these high-temperature polymers need to be better defined, and the best binder polymer needs to be established. One area that should be investigated is the use of inorganic or organic polymers such as can be produced using silane chemistry. Because a portion of such polymers will have inorganic features and another part will be organic in nature, this should lead to enhanced adhesion being possible between the inorganic particles and the polymer separator substrate.

Thus, for the fundamental studies of ceramic-coated battery separators, certain advantages of actual separator features have been determined and measured. Among those features are both the thermal stability and wettability of the separator to the electrolyte. Different levels of both of these effects have been measured and actual improvements that have been observed depend on the nature of the formulation used in the coating process.

The translation of the improved separator features into battery safety requires further work. Results of abuse testing, such as nail penetration or ball crush tests, need to be documented to show measured and quantifiable improvements in battery safety that are obtained with a certain level of thermal stability improvement. Establishment of that correlation could lead to further areas of work that can be done to optimize the ceramic coating approach completely.

Ceramic coating patents

Many patents involve the development of ceramic-coated battery separators. Therefore, the intellectual property situation is not clear. This section of the chapter summarizes main findings from various patents and attempts to define the unique features of inventions that are described in each of them.

The first patent to be discussed is Celgard RE47,520 formerly US Patent 6432586 [24]. The patent discusses a battery separator that includes at least one ceramic composite layer and at least one polymeric microporous layer. The ceramic composite layer includes a mixture of inorganic particles and a matrix material. The structure of the ceramic composite layer is designed to block the growth of dendrites and prevent electronic shorting. On the other hand, the polymeric layer is adapted to block the flow of ions between the anode and cathode in the event of a thermal runaway situation.

In the patent, the ceramic composite layer is described as being nonporous. The preferable composition of the ceramic composite layer contains 40%–60% inorganic particles by weight. Preferred inorganic particles are described as SiO_2, Al_2O_3 and $CaCO_3$, with a preferred particle size of 0.01–2 μm.

No specific properties of the coated separators are provided in the patent. In addition, the performance of batteries that contain the separators is not explicitly shown. Thus, it is not clear what level of safety the proposed coated separators actually provide to the lithium-ion batteries.

Another patent, US Patent 9711771 was filed by Celgard [25]. That patent addresses the point that some large ceramic particles are difficult to load and disperse into the polymeric layer of the coating layer because of a difference in surface energy between the particles and the polymeric resin material. The patent uses particles with an average particle size of less than 1 μm. The particles have a coating that contains a molecule with a reactive end that can bind to the surface of the particle and a nonpolar end that can bind to the polymeric material. The surface coating alters the surface energy of the particle to be similar to that of the polymeric material. This allows the nanoparticles to be better mixed with the polymeric material.

LG Chem has also been a key player in producing ceramic-coated separators. For example, US Patent 7638241 [26] discusses an organic—inorganic composite separator with a porous substrate as well as a porous coating layer consisting of a mixture of inorganic particles and a binder polymer. The coating layer is specified as having a content ratio of binder polymer—inorganic particles that is higher in the surface region than it is inside the coating layer. The use of a separator that contains such a layer is described as having improved lamination features for battery electrodes. Stability and performance of the lithium-ion battery can be improved, because detachment from the coating layer may be reduced in assembling the battery.

US Patent 7662517 [27] specifically discusses the use of a polyolefin-based substrate separator. The polyolefin substrate is coated with a mixture of inorganic particles and a binder polymer. The inorganic particles are interconnected and are fixed by the presence of the binder polymer. Interstitial volumes among the inorganic particles form a pore structure in the coating layer. The inorganic particles that are most useful in the invention have dielectric constant values of 5 or more. Batteries that contain the organic—inorganic composite separator have improved thermal safety, electrochemical safety, and quality.

US Patent 7709152 [28] further discusses an organic—inorganic composite separator in which the porous coating also contains a mixture of inorganic particles and a polymer. That layer is coated onto the surface of a porous substrate. The work attempts to optimize the inorganic particle—binder formulation. Previous work required a certain amount of inorganic particles to restrain the thermal shrinkage of the polyolefin porous substrate. However, as the content of inorganic particles increased, the content of the binder polymer was relatively decreased. This can lead to at least two problems in the battery.

First, owing to the stress generated in battery assembly, such as winding, the inorganic particles may be extracted from the coating layer. The extracted inorganic particles can act as a local defect in the battery. This have a bad influence on the stability of the battery.

Second, adhesion between the coating layer and the porous polyolefin substrate is weakened. This leads to an ability of the porous coating layer to restrain thermal shrinkage of the polyolefin substrate from deteriorating. Because of that effect, it is difficult to prevent an electric short circuit between the electrodes, especially when overheating occurs in the battery.

Those issues are addressed in the patent through the use of a mixture of two binder polymers with have completely different hydrophilic properties. This allows for the formation of a polymer blend with controlled hydrophilic properties. Using such a blend, a synergistic effect in the improvement of thermal stability of the organic–inorganic composite membrane can be obtained. Specifically, thermal shrinkage values of 50% or below are observed in the machine direction or transverse direction at 150°C. In addition, the coating layer has excellent peeling resistance, as judged by a peeling force of at least 5 gf/cm.

These properties are achieved through a blend of two binder polymers, one of which has a contact angle of 90 to 110 degrees and a second binder polymer with a contact angle of 20 to 40 degrees. Some examples of the first binder polymer are PVDF, PMMA, and polyimides. The second binder polymer needs to have at least one polar group selected from a hydroxyl group (−OH), carboxyl group (−COOH), maleic anhydride group (−COOOC−), sulfonate group (−SO_3H), or pyrrolidone group (−NCO−). A particularly useful second binder polymer seems to be cyanoethyl polyvinyl alcohol.

LG Chem extended those studies to include a binder polymer that is a copolymer [29]. The copolymer needs to consist of (1) a first monomer unit with a contact angle between 0 and 49 degrees, and (2) a second monomer unit with a contact angle of 50 to 130 degrees. The copolymer can be either a random or block copolymer.

The first monomer has a relatively greater hydrophilic property than the second monomer unit and thus contributes to adhesion among the filler particles. Also, the second monomer unit has a relatively greater hydrophilic character than the first monomer unit. Thus, the presence of the second monomer unit contributes to an improvement in adhesion between the filler particles and the separator substrate. Thus, when the copolymer is used as a binder polymer in the ceramic coating layer, the peeling resistance of the coating layer can be increased compared with the use of a conventional binder polymer. Therefore, it is possible to lower the content of the binder polymer and thus increase the content of the filler particles. The thermal shrinkage of the porous substrate may be further controlled. Because the adhesive strength between the porous coating and the porous substrate is reasonable, the thermal shrinkage of the substrate can be controlled if the battery experiences overheating.

LG Chem expanded on this work by developing a porous coating layer with a shutdown function that exhibits appropriate electrical conductivity [30]. This makes the separator suitable for use in high-capacity batteries. Such batteries have many applications, including computers and electric vehicles.

The goal is achieved through the use of a nonwoven fabric substrate and a porous coating layer composed of a mixture of inorganic filler particles and a binder polymer. The filler particles are composed of a mixture of conductive particles and a low—melting point resin that has a melting point lower than that of the nonwoven fabric substrate. The presence of the conductive particles means that the porous coating layer exhibits appropriate electrical conductivity. The separator is suitable for use in a high-capacity battery. Furthermore, overheating of the battery expands the low—melting point resin that is mixed with the conductive particles. In particular, when the thermal runaway occurs in a battery, the low—melting point resin melts and closes the pores of the porous coating layer to suppress electrochemical reactions from proceeding further.

The conductive particles used in this invention should be both electrically conductive and thermally stable. Carbon materials such as natural graphite, artificial graphite, or carbon black are proposed as electrically conductive particles that can be used. The example in the patent uses carbon black as the conductive particle.

The low—melting point resin used with conductive particles needs to have a lower melting point than that of the nonwoven fabric substrate. The low—melting point resin melts before thermal runaway occurs in the battery. This causes shutdown in the separator. A polyolefin resin such as PE or an ethylene copolymer, which has a melting point of 80—130°C, is preferable to provide the shutdown function to the separator.

LG Chem has provided a patent that discusses specific features desirable for the separator in high-power electrochemical devices [31]. The patent involves the use of a monolayer-type polyolefin-based microporous film with a porosity of 40%–60% and an average pore diameter of 60 nm or less. The porous coating layer is formed on at least one surface of the microporous film and is made of a mixture of inorganic particles and a binder polymer.

The separator controls a short circuit between the anode and cathode. The separator, with the inorganic particles in the porous coating layer, can reduce overheating in the battery. In addition, the presence of the porous coating layer controls shrinkage so that it can be further lowered.

Inorganic particles in the porous coating layer are connected to each other by means of the binder polymer. Pores in the coating layer are formed among the inorganic particles, which serve as a type of spacer that keeps a physical form to the porous coating layer.

It is preferable for the inorganic particles to include high dielectric inorganic particles with a dielectric constant of 5 or above. In other words, the particles might have the ability to transfer lithium ions. The inorganic particles may have the function of carrying lithium ions without storing lithium.

US Patent 9559345 [32] discloses a battery separator that includes a nonwoven fabric substrate, fine thermoplastic powder located inside the pores of the nonwoven fabric substrate, and a porous coating layer dispersed onto at least one surface of the nonwoven fabric substrate. The thermoplastic powder has an average diameter smaller than that of the pores. In addition, it has a melting point lower than the melting or decomposition point of the nonwoven fabric substrate. The porous coating layer includes a mixture of inorganic particles and a binder polymer whose melting point is higher than either the melting or decomposition point of the fine thermoplastic powder. The inorganic particles in the coating layer are connected to each other by the binder polymer. The pores in the coating layer are formed by interstitial volumes between the inorganic particles.

Filling the large pores of the nonwoven fabric with the thermoplastic powder makes the porous coating layer uniform. This maintains the uniformity of the porous coating layer even when the separator undergoes stress during battery fabrication. The occurrence of charging failure and leakage can be prevented.

US Patent 9666849 to LG Chem uses a binder polymer that is a copolymer containing either an amine or an amide as well as a second monomer unit of methacrylate. The coating layer has a high packing density and good adhesion strength with a porous substrate. These effects result in decreased resistance, which may contribute to an increased capacity for the lithium-ion battery. In addition, the separator has increased resistance against thermal and mechanical impact. This can prevent detachment of the inorganic particles in the porous coating layer.

US Patent 9698396 [33] discloses a separator that possesses a dual porous coating layer of inorganic particles with different surface features. The coated separator is produced by forming two different slurries and hence, two separate porous coating layers. The separator that is produced has a uniform dispersion in the coating layer. The separator adsorbs excess metal ions that are generated when the battery is outside a normal temperature range. That effect contributes to the safety of the battery.

US Patent 10002719 [34] discloses a coated separator that uses a binder polymer mixture consisting of two PVDF-HFP copolymers that differ in HFP by 3 wt or higher. If the two copolymers in the binder mixture have differences outside the preferred composition ranges, the binder layer that is formed will have poor adhesion with the electrodes. The composition must also be optimized so that the binder can effectively connect or immobilize the inorganic particles on the porous substrate. In addition, the binder layer can be too thick and the pores in the porous coating layer can become plugged. This can lead to a decrease in the Gurley air permeability, which suggests deterioration in battery performance.

US Patent 10158111 [35] discusses the use of an aqueous slurry that contains pulverized inorganic particles, a binder polymer, and an aqueous medium. The binder polymer is either a styrene-butadiene rubber-based polymer or an acrylate-based polymer. The use of the aqueous slurry is safer, more ecological friendly, and economically more efficient than an organic-based slurry.

The surface tension of the aqueous slurry is an important factor for guaranteeing good wettability on the porous polymer substrate. To achieve the needed surface tension level, an additive can be introduced into the aqueous slurry. Various additives including siloxane and modified castor oils are quoted as being useful in that regard.

The aqueous medium used in the aqueous slurry may be water or a mixture of alcohol and water. Examples of the alcohol that may be used include methanol, ethanol, and propanol. No specific alcohol is described as being preferred.

US Patent 10541399 [36] discloses a coated battery separator that uses polymer particles as a binder compared with a solvent-type binder resin that uses organic solvents for implementation and use. The separator that is produced has improved air permeability, which can easily be controlled by monitoring the content of the binder and/or thickener. The binder polymer particles in the organic–inorganic composite porous layer do not penetrate the pores of the separator substrate; this is the reason for the outstanding air permeability features. This is a different situation than that which occurs when a solvent-type binder solution is used when penetration into the pores of the porous substrate occurs and the result is a reduction in the porosity of the separator.

Polymer particles with a particle size larger than the pore size of the porous substrate do not raise the problem of blocked pores in the porous substrate. The porosity of the porous substrate is unaffected. Also, because the polymer particles are distributed in the composite porous layer in a mixed state with the inorganic particles, there is an increase in the bond between the inorganic particles compared with a solvent-type binder.

The final LG Chem patent to be discussed is US Patent 10756319 [37]. The invention involves coating a slurry containing inorganic particles and a first binder polymer dissolved in a first solvent. Then, a second binder polymer that is dissolved in a second solvent is coated on the slurry. The first and second solvents are dried simultaneously to form a porous polymer outer layer of the second binder polymer and an organic–inorganic composite inner layer.

The porous polymer outer layer that is formed on the surface of the organic–inorganic composite inner layer improves the ability of the separator to bond to the electrodes and allows for easy lamination. Second, the porous polymer layer serves as a mesh to prevent inorganic particles from detaching from the organic–inorganic composite inner layer. This prevents subsequent problems that may occur because of detachment of the inorganic particles. Also, the presence of the porous polymer outer layer prevents content loss of the inorganic particles in the organic–inorganic inner layer, thus improving the stability of the separator. This is important, for example, to have good bonding of the porous coating layer with the electrodes during battery assembly, such as during the winding process.

Much of the work by LG Chem focused on using different binder polymers in the ceramic coating formulation. Different binder polymers and combinations of them have been used to produce the ceramic coating. The motivation for much of the work has been to improve the thermal stability of the battery separator. The improvement is related to enhancements in the safety of the battery.

Although much work has been done on using various binder polymers, a final conclusion has not been reached as to which binder polymer or combination of polymers is most effective for use in the ceramic coating. It is not clear what structural features of the binder polymers are most effective. In addition, it is not obvious whether a single binder polymer is most effective for all inorganic particles. It is possible that each inorganic filler may have an optimal binder polymer. The definition of features of the various binder polymers that permit them to work effectively in ceramic coating formulations needs to be established.

Along those lines, the optimal inorganic filler has not yet been defined. There has been discussion in several of the LG Chem patents of using filler particles with a dielectric constant of 5 or more. This is a first step in defining the most effective inorganic filler particles to use, but more work needs to be done to define features that provide the greatest improvement in battery safety.

Other organizations have pursued patents in the area of ceramic-coated battery separators. For example, SK Innovation [38] disclosed a composite separator that includes a coating layer with a thermally cured aqueous slurry containing inorganic particles, binder particles, a second binder, and a thermal curing agent. The binder particles are made from a copolymer with specific monomer units such as acrylic monomers and acrylamide-based monomers. The composite separator that is produced has thermal shrinkage of 160°C of 3% or less and a rupture temperature of 400°C or more. These effects lead to an improvement in thermal stability and an increase in battery life.

The presence of the second binder material in the formulation means that the melt fracture temperature is improved. For example, with the inclusion of the second binder, melt fracture temperatures of 400°C or more can be achieved. This is believed to be because the presence of the second binder polymer improves the curing density; hence, a coating layer of excellent strength is provided.

Applied Materials provided a patent that discloses uniform ceramic coatings with controlled thicknesses for separators [39]. The approach has less waste and higher yields than other ceramic coating technologies. The technology involves the use of a vapor deposition process to deposit a ceramic film on a porous polyolefin separator. The resulting coated structure has increased thermal stability and increased wettability. The coated separator has enhanced ionic conductivity, which is believed to be caused by liquid–surface interactions between inorganic particles in the pores of the separator material. This allows the entire separator structure to be reduced in thickness below 10 μm. This is necessary for the production of high-power batteries for use in electric vehicles.

GM Global Technology Operations [40] produced separators formed predominantly from heat-resistant particles. The collection of heat-resistant particles is held together by a relatively small amount of a porous inert polymer material, which acts as a binder to hold the heat-resistant particles together.

The heat-resistant particles may be ceramic particles, cross-linked thermoset polymer particles, or engineering thermoplastic particles. Some examples of suitable ceramic particles include alumina, titania, and silica. Cross-linked thermoset particles include those of polyurethanes and epoxy resins. Suitable engineering thermoplastic polymer particles include those of polyimide, polyphenylene sulfide, polyether ether ketone, and polysulfone. The high content of the heat-resistant particles provides the separator with robust thermal stability well above 130°C. The porous polymer layer is formed through a phase separation process.

All of the patents discussed for ceramic-coated battery separators focused on improving the thermal stability of the separator. That effect enhances the safety of lithium-ion batteries. This is particularly important for large batteries used in electric vehicles.

More work needs to be done to explain and understand the observed effects. This work needs to be a combination of both fundamental and patent studies to optimize the performance of batteries that contain ceramic-coated separators. Once a formulation is identified that shows promise in batteries, fundamental studies should be undergone to understand the features of the formulation that provide the positive aspects. Questions that need to be addressed include those regarding the inorganic particle type and size. Also, for a particular type of inorganic particle, the most effective binder polymer needs to be defined. Answers to those issues will optimize ceramic coating technology to enhance the performance of batteries to the highest possible level.

Commercial activity

Miltec is a company that developed a revolutionary patented technology that gives lithium-ion battery manufacturers the ability to produce ceramic coatings on lithium-ion battery separators [41]. The process involves an energy-saving UV coating process that nearly eliminates the need for a drying oven.

Separators can be coated at 120 m/min. The ceramic coating units occupy a small space of 1 m × 2 m. The UV binders that are used are electrochemically stable and have been tested in batteries that operate at up to 4.8 V. The cycling stability of the UV chemistry is equivalent to what is observed with the use of PVDF.

Celgard offers ceramic coating options at several different coating thicknesses. These include options from 2 to 5 μm thick. The ceramic coating is described as providing enhanced short prevention and excellent structural integrity at high temperatures. A picture of the Celgard base film coated with a ceramic layer is shown in Fig. 7.1.

Ube Maxell Co., Ltd. signed a patent license agreement with LG Chem, Ltd. for coated separator film technology owned by LG Chem [42]. Under the terms of the agreement, Ube Maxell acquires patent use rights for the ceramic coating of separator substrates. The patent granted to Ube Maxell is

Figure 7.1 Celgard base film with a ceramic coating layer.

for LG Chem's safety-reinforced separator technology. That technology uses a ceramic coating on separator substrates to enhance thermal deformation resistance and the mechanical strength of the separators.

Optodot developed ceramic-coated separators based on using boehmite in place of alumina [43]. Boehmite has a 25% lower density than alumina and provides a lower cost and lower coating weight. The coating provides greater mix stability and ability to coat. In addition, boehmite is much less abrasive than is alumina, which provides for the extended life of coating and slitting and process equipment. Optodot entered into several licenses under its patent portfolio that cover separators that contain boehmite.

Faustel provides equipment to perform battery separator coating [44]. The pre-engineered systems have customizable features to allow for any separator coating process. They are available in both single- and double-pass configurations. They possess in-line coat weight measurement capabilities as well as the availability to have solvent recovery and distillation features. A pilot coating line is also available to predict coating features in various operations before scaling up to production-level equipment. In addition, coating formulations can be evaluated quickly as a way to optimize the final separator performance.

As with the base separators, there is increased activity in the commercialization of ceramic-coated battery separators. The initial coating efforts were performed by companies with specific experience with film coating. Often, that experience involved coating films for use in applications other than lithium-ion batteries. The processes developed for these other films needed to be modified to adapt them to battery separators. As those modifications occurred, some of the knowledge was extended by the producers of the separator substrates. That has allowed some of the producers of the substrate materials to perform the coating production work themselves.

This transfer of the technology has meant that the cost of the coating process was reduced. This is because costs in terms of transferring the films between different locations can be reduced or eliminated. The trend of more substrate producers performing the ceramic coating process is likely to continue.

In addition, it is expected that more organizations will enter the business of ceramic coating battery separators. The quantities of film that can be coated for use in electric vehicles and other applications is large enough to encourage new players on the market. Furthermore, there are opportunities for evaluating new formulations and new materials. In that sense, pilot line evaluations are critical to the eventual scale-up of coating formulations to production-level equipment.

Conclusions

This chapter has summarized both fundamental and applied studies involved in producing and testing ceramic-coated separators. Much work has been done in both the academic world and industrial areas. Many articles have been written and numerous patents have been issued on the topic.

Most work has focused on increasing the thermal stability of the separator by using a ceramic coating on the separator substrate. An increase in thermal stability means that thermal shrinkage of the separator is decreased at elevated temperatures. Thus, electric short circuits will be prevented in the battery and the potential for fires and explosions will be reduced. This is particularly important for large batteries that are used in electric vehicle applications.

More work needs to be done to better define the optimal formulation for use in ceramic coating of battery separators. The best inorganic filler to obtain the highest increase in battery safety needs to be defined. The chemical structure needs to be finalized, but also within a specific chemical structure, the most useful particle size needs to be determined. For the use of a specific inorganic filler material, the features of the most efficient binder polymer need to be established. Much work has already been done with various binders, but additional explanations and rationale need to be provided to understand the selection process better for the appropriate binder to use in the ceramic coating formulation.

That understanding can likely best be achieved through a collaboration among appropriate academic and industrial organizations. Once the ceramic coating formulation is defined, fundamental studies need to be performed to understand the reasons for the observed effect. This is all part of a need to establish structure—property relations for battery performance and its relation to the separator structure. This is particularly significant as it relates to battery safety testing, such as nail penetration or ball crush testing. This is because increases in thermal stability for the separator that are provided by the ceramic coating are related to an enhancement in battery safety. A quantitative relationship between the features should be established to understand the correlation between them.

Finally, further advantages of applying a ceramic coating onto battery separators using different techniques need to be expanded. Much work thus far has been done using the dip-coating method of applying the coating. However, approaches that involve both the surface treatment of the inorganic particles and the separator substrate have yielded promising results in terms of improving adhesion between the particles and the porous substrate. This assists in eliminating the detachment of particles from the separator, and thus makes the battery assembly process easier to perform without concerns regarding that issue.

Ceramic-coated battery separators offer potential advantages for lithium-ion batteries. Those advantages are currently being exploited, and future prospects appear to be high. However, work need to continue to define actual structural features of the separator that provide the largest advantages. That definition will allow for the most effective use of the technology in future and developing applications.

References

[1] Song, et al., Electrochim. Acta 85 (2012) 524.
[2] www.mirvec-coating.com/technology/microgravure.
[3] L. Yu, Y. Jin, Y.S. Lin, RSC Adv. 46 (2016).
[4] L. Yu, J. Mao, Y. Jin, J. Lin, Front. Chem. Sci. Eng. 11 (3) (2017) 346.
[5] C. Shi, P. Zhang, L. Chen, P. Yang, J. Zhao, J. Power Sources 270 (2014) 547.
[6] H. Liu, J. Xu, B. Guo, X. He, Ceram. Int. 40 (2014) 14105.
[7] Y.M. An, T.H. Kim, D.R. Chang, J. Solid State Electrochem. 18 (2014) 1807.
[8] C. Shi, et al., J. Membr. Sci. 517 (2016) 91.
[9] Lee, et al., J. Power Sources 294 (2015) 537.
[10] P. Yang, P. Zhang, C. Shi, L. Chen, J. Dai, J. Zhao, J. Membr. Sci. 474 (2015) 148.
[11] D. Fu, B. Luan, S. Argue, M. Bureau, I. Davidson, J. Power Sources 206 (2012) 325.
[12] K. Prasanna, C.S. Kim, C.W. Lee, Mater. Chem. Phys. 146 (2014) 545.
[13] R.S. Juang, C.T. Hsieh, P.A. Chen, Y.F. Chen, J. Power Sources 286 (2015) 526.
[14] K. Prasanna, T. Subburaj, W.J. Lee, C.W. Lee, Electrochim. Acta 137 (2014) 273.
[15] D. Yeon, Y. Lee, H.M. Lee, Y.M. Lee, Electrochim. Acta 157 (2015) 282.
[16] H. Jeon, S.Y. Jin, W.H. Park, H. Lee, T.H. Kim, M.H. Ryou, Y.M. Lee, Electrochim. Acta 212 (2016) 649.
[17] X. Zhu, X. Ziang, X. Ali, H. Yang, Y. Cao, J. Membr. Sci. 504 (2016) 97.
[18] X. Jiang, X. Zhu, X. Ali, H. Yang, Y. Cao, ACS Appl. Mater. Interfaces 9 (2017) 25970.
[19] Y. Choi, J. Kim, J. Moon, J. Jeong, J.H. Park, Appl. Surf. Sci. 444 (2018) 339.
[20] M. Liu, P. Zhang, L. Gou, Z. Hou, B. Huang, Mater. Lett. 208 (2017) 798.
[21] W. Zheng, Y. Zhu, B. Na, R. Lu, H. Liu, W. Li, H. Zhou, Compos. Sci. Technol. (2017) 144−178.
[22] Wu, et al., RSC Adv. 7 (2017) 2440.
[23] W. Li, X. Li, A. Yuan, X. Xie, B. Xiu, Ionics 22 (2016) 2143.
[24] RE47,520, Issued to Celgard, LLC, July 16, 2019.
[25] U.S. Patent 9,711,771, Issued to Celgard, LLC, July 18, 2017.

[26] U.S. Patent 7,638,241, Issued to LG Chem. Ltd.
[27] U.S. Patent 7,662,517, Assigned to LG Chem, Ltd, Febuary 16, 2010.
[28] U.S. Patent 7,709,152, Assigned to LG Chem, Ltd, May 4, 2020.
[29] U.S. Patent 8,338,018, Assigned to LG Chem, Ltd, December 25, 2010.
[30] U.S. Patent 8,703,323, Assigned to LG Chem, Ltd, April 22, 2014.
[31] U.S. Patent 9,142,819, Assigned to LG Chem, Ltd, September 22, 2015.
[32] U.S. Patent 9,559,345, Assigned to LG Chem, Ltd, January 31, 2017.
[33] U.S. Patent 9,698,396, Assigned to LG Chem, Ltd, July 14, 2017.
[34] U.S. Patent 10,002,719, Assigned to LG Chem, Ltd, June 19, 2018.
[35] U.S. Patent 10,158,111, Assigned to LG Chem, Ltd, December 18, 2018.
[36] U.S. Patent 10,541,399, Assigned to LG Chem, Ltd, January 21, 2020.
[37] U.S. Patent 10,756,319, Assigned to LG Chem, Ltd, August 25, 2020.
[38] U.S. Patent 10,756,322, Assigned to SK Innovation Co., Ltd, August 25, 2020.
[39] U.S. Patent 10,756,321, Assigned to Applied Materials, Inc, August 25, 2020.
[40] U.S. Patent Application 20120308872, Assigned to GM Global Technology Operations, LLC, December 6, 2012.
[41] miltec.com/en/technology/battery/uv-ceramic-coated-separators/.
[42] ube-ind.co.jp/ube/en/news/2014/20140516 - 01.html.
[43] optodot.c om/lithium - battery - technology#licensing.
[44] faustel.com/converting-equipment/specialized-solutions/battery-separator coating/.

Further reading

[1] U.S. Patent 9,666,849, Assigned to LG Chem, Ltd, May 30, 2017.

CHAPTER 8

Unmet needs for battery separators

Introduction

Uses for lithium-ion batteries are expanding each year. These uses occur in the consumer electronics field, where new devices are constantly being developed. They are also happening in the electric vehicle market as new types and models of vehicles are being introduced. Finally, there are novel uses for lithium-ion batteries in emerging areas such as power grids and power tools.

These emerging uses mean that a reexamination of battery components, including the separator, is necessary. The separator structure used in other applications will need to be suitable in these areas. This has to be determined and verified by testing and evaluating current separator structures in the new uses. If it is determined that the separator structure is inappropriate, existing structures will need to be adjusted or modified or completely new structures will need to be defined.

Because of these emerging uses, there are several unmet needs for current battery separators. It is possible that these needs can be satisfied using current wet and dry process separator technologies with the necessary process adjustments. On the other hand, new process and materials approaches may need to be established to meet these uses. To address those questions, more research and development work is required. Those efforts will need combined materials suppliers and battery producers to optimize results fully in terms of battery performance.

This chapter will address some current unmet needs for lithium-ion battery separators. These needs are based on important criteria for next-generation batteries. Such batteries will have applications in electric vehicles. In addition, new areas for lithium-ion batteries such as electric grids can be envisioned. These uses will likely have novel requirements for separators that may not be met by current separator materials. Thus, new separator materials with new requirements may be necessary for these applications.

Polymer-Based Separators for Lithium-Ion Batteries
ISBN 978-0-12-820120-6
https://doi.org/10.1016/B978-0-12-820120-6.00010-6

In addition, the development of new materials for lithium-ion batteries is leading to the need for new separator materials. For example, ionic liquids (ILs) are widely being studied as a safer alternative electrolyte material for use in lithium-ion batteries. IL electrolytes have poor wetting with commercial separators. The wetting issues most likely result from poor compatibility with the separator, not poor electrode compatibility. A separator that is compatible with IL electrolytes needs to be developed before the electrolytes can be used in commercial lithium-ion batteries. This topic will be discussed in much greater detail in this chapter.

Several unmet needs for battery separators for use in lithium-ion batteries will be discussed: (1) thinner separators, (2) high-temperature separators, (3) separators with high porosity, and (4) electrolyte-absorbing separators. Specific needs in each area will be highlighted. In addition, current research efforts in each topic will be discussed and areas that need further work and study will be pointed out.

Thinner separators

Most commercial separators are currently on the order of 12 μm thick or greater. These products were developed for consumer electronics applications. For example, Entek provides a series of products based on wet process technology with have typical thicknesses ranging from 12 to 20 μm [1].

However, thin battery separators provide several advantages that can improve the battery's performance and design. First, thinner separators attain higher rate capabilities. This suggests that the battery is able to produce higher current densities across the separator because the electrical resistance of the separator can be reduced by reducing the separator's thickness. The higher rate capabilities are important when quick power is needed, as in the acceleration of an electric vehicle.

Second, thinner separators allow for the use of thinner electrodes. This translates into the battery attaining better electrode use and electrode cycling. Electrode use means that smaller batteries with equivalent power levels can be produced. Better electrode cycling means a greater number of rechargings or cycles over the battery's life.

Finally, thinner separators allows for a greater choice in the design of the separator. Various thin separators can be combined to custom design a separator for a particular battery. The separator structure can be tailored to meet specific needs for a type of battery. This further indicates that the properties of the separator can be adjusted as needed.

A primary motivations for developing thin separators is interest in higher-power and higher-energy batteries for use in electric vehicles. One way to achieve that goal is to employ different electrode materials than are currently used. However, there is a limit to levels of increases in power that can be achieved with this approach, because only certain materials can be used in electrodes.

However, the use of thin separators is an alternative approach to achieve high-power batteries. The use of thin separators means that more active electrode material can be used in the battery. Owing to the presence of the increased amount of electrode material, the produced battery can have an increased power and energy level. This approach means that the manufactured battery can be the same size as current batteries. Devices that use the batteries do not need to be reconfigured to deal with larger batteries. This is an important feature of using thin separators to produce high-power batteries.

Much work on thinner separators has been done using the wet process approach [1]. This is primarily because at the same thickness, wet process separators generally have higher puncture strength values than do dry process separators. This is because wet process separators are biaxially oriented compared with dry process separators that are uniaxially oriented. Higher puncture strength values translate into higher puncture strengths at low thicknesses. This can be observed because of the correlation between the film thickness and the puncture strength [2]. Puncture strength is necessary for the separator to provide safety to lithium-ion batteries.

In wet process technology, to produce films with lower thicknesses, it is necessary to cast the films thinner than when making thicker films. This may make it necessary to adjust the processing parameters required to produce the cast film. For example, thinner separators will require different extruder speeds. This can lead to different melt viscosities that may result in handling problems with the thin cast films.

Furthermore, the production of thinner cast films may require modification of the die used for the extrusion process. This is because the production of thin cast films will require tighter tolerance levels in terms of the actual standard deviation that is acceptable to produce adequate film products. Deviations in the cast film thickness can be directly translated into thickness variations in the final film products. Thickness variations that are present in the original cast film will not be resolved by further processing such as film stretching.

The remainder of the details on the wet process separator are the same as those presented in Chapter 5. The ideal stretch ratios for the stretching process will likely be the same as those that were previously discussed. However, because thinner films are being produced, it is anticipated that handling of the final film products will need to be dealt with carefully. This is necessary to prevent film breaks that reduce the overall product yield. Such a decrease in yield could negatively affect the cost associated with producing thin separators.

Thus far, it has been possible to produce separators as thin as 9 μm. There is also continued interest in producing even thinner separators. Work has been done to produce separators as thin as 5 μm [1]. However, the puncture strength of these separators is low. This translates into concerns about safety provided to the battery using these thin separators.

One way to deal with the low puncture strength of thin separators made by the wet process is employ an ultrahigh-molecular weight polyethylene (PE) (UHMWPE) material in the battery separator. For example, Celanese provides several UHMWPE polymers with molecular weights of 5 million or more [3]. The use of such a high-molecular weight polymer offsets the fact that thinner separators have lower mechanical properties such as puncture strength.

Celgard has addressed the need to produce dry process separators that are relatively thin [4]. This patent discloses a novel or improved battery separator made from a single or a multilayer microporous polyolefin membrane with a thickness of 12 μm or less. In particular, shutdown separators are discussed. It is proposed that the PE density at the PE-polypropylene (PP) interface in the shutdown membrane is more important than the overall amount of PE material that is present in the shutdown layer in the separator. This is different from the usual assertion stating that the shutdown behavior is linked to the overall thickness of the PE shutdown layer. The patent claims that it is possible to increase the density of the PP-PE surface by controlling the extrusion, stretching strain rates, and relaxation of the separator. However, it does not provide complete specific guidance for process parameters that should be used to optimize the PE density at the interface. Nevertheless, control of the PE density allows for complete shutdown of the separator to be achieved with a much thinner PE section in the separator than was present in previous separator structures.

The invention of this patent increases the separator's tortuosity by developing an interface between the layers of the multilayer separator.

The interface has a microporous structure defined by its own pore size, porosity, thickness, and tortuosity. The tortuous porous network functions to reduce or prevent the growth of dendrites through the separator membrane. This can prevent the occurrence of internal short circuits resulting from dendrite growth. Also, it can provide the battery with an enhanced cycle life.

More work needs to be done to define the relation between puncture strength and thickness quantitatively for these thin separators. Because puncture strength is one measure of safety that the separator can provide to overall battery safety, this is a necessary relation to understand safety issues that may exist with the use of these thin separators. Safety is a primary concern here, and it is important that it be addressed.

This relatively low puncture strength issue can be addressed in wet process separators through the use of UHMWPE polymers with the highest molecular weight possible. Such polymers have higher mechanical properties than do other PE materials, which this should increase the puncture strength of the thin separators. In addition, there may be additives to the PE formulation to increase the puncture strength of the separators.

PE polymers with increased molecular weight have increased processing issues [5]. This largely manifests in having film breaks during the stretching step of production. The overall yield for the production might be as high as when low-molecular weight PEs are used. The economic trade-off of this effect needs to be evaluated in terms of possible advantages provided by the high puncture strength that can be obtained.

The other issue that using UHMWPE materials raises is related to the shutdown features of the separators. Because of the melt viscosity of the UHMWPE polymers, flow is a slow event at the melting point of 135°C. Thus, the flow of PE that is responsible for closing the pores and for the subsequent separator shutdown does not occur.

To accomplish that effect, a lower-molecular weight PE polymer or a different type of low melting PE needs to be added. Its presence will naturally lower the strength of the separator made from the formulation. The actual amount that the strength will be lowered depends on the amount of the other polymer in the formulation. Thus, to minimize this effect, the lowest amount of the other polymer is desirable. On the other hand, a sufficient amount of the other PE polymer needs to be present to guarantee that the shutdown effect occurs. The balance of the two effects needs to be realized to have the optimal formulation in the battery separator.

For thin dry process separators, more work needs to be done to define the separator properties that occur in the separators that are produced. Also, it needs to be determined whether the observed properties lead to handling problems with the separator. It is not clear whether establishing the increased PE density at the PE-PP interface has a significant effect on the overall property profile of the separator. This question needs to be studied in more detail to implement results that are obtained.

High-temperature separators

Much work to develop high-temperature separators was been discussed in Chapters 6 and 7 of this book. However, work continues on this topic because of the challenges present in the approaches that were developed thus far. This section of the chapter will summarize those efforts and will discuss some areas for additional work.

First, the major driving force for these developments is the electric vehicle market. The larger batteries used there require protection from thermal runaway effects. The separator contributes to this effect by keeping the electrodes apart at elevated temperatures.

There is no currently accepted requirement for a high-temperature separator. In addition, no standard test has been developed to define a high-temperature separator. Specifically, no temperature is generally agreed upon as defining a high-temperature separator.

For example, one discussion [6] states that an evolving separator demand is for higher-temperature stability in the range 200—250°C. The separators are stated as needing to retain sufficient dimensional stability within that region of temperatures. However, no specific criteria in terms of acceptable shrinkage values at these elevated temperatures are provided in this discussion. Finally, no explanation is provided as to why that particular temperature range for dimensional stability is required.

On the other hand, USABC defined advanced battery separators as having an high temperature melt integrity (HTMI) >150°C [7]. This is certainly different from the previously described report. Also, there does not appear to be a rationale provided for this temperature specification. However, the poly(etherimides) described in the article are touted as new material candidates for high-performance separators.

In addition to providing high-temperature stability to the battery, there is continued interest in having those separators provide specific beneficial functions. A review of functional membranes for next-generation lithium batteries has been provided [8]. Such functional separators would have selected permeability features and could retard the migration of a specific species from one electrode to another to achieve stable cycling operation.

No one single test can be used to define the high-temperature performance of battery separators. One test that can be used is a determination of the rupture temperature of the separator. A typical thermomechanical analysis (TMA) test to define rupture temperatures is shown in Fig. 8.1.

In that experiment, dimensional changes in the separator are measured as a function of temperature. The temperature at which the separator completely breaks with the specific load is defined as the rupture temperature of the separator. To have a separator with thermal stability requires a high rupture temperature.

Three approaches are being considered to improve the high-temperature behavior of battery separators. First, new polymers and new polymer formulations are being used in battery separators. An example of that was briefly noted in l'Abee et al. [7]. In that article, high temperature–resistant lithium-ion battery separators based on novel polyetherimides and made through a phase inversion process are described. Dynamic mechanical analysis demonstrated dimensional stability for separators up to 220°C.

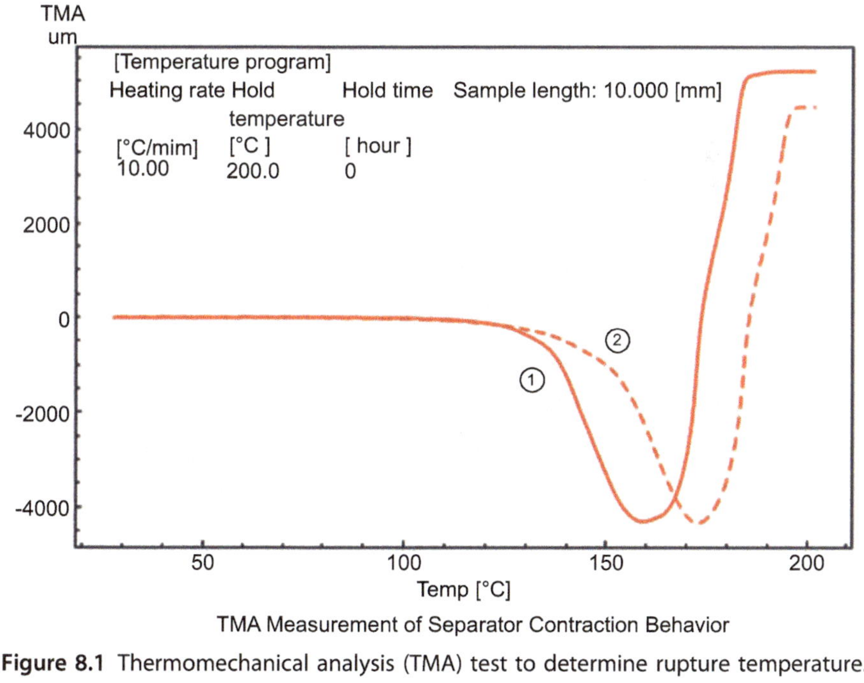

Figure 8.1 Thermomechanical analysis (TMA) test to determine rupture temperature.

Poly(ether-ether-ketone) (PEEK) membranes were made by a phase inversion process [9]. A sponge-like porous membrane was produced using this technology. The PEEK membrane showed no fusion deformation even at 350°C, and had excellent thermal stability.

The other class of polymers that has attracted much research attention in high-temperature separators is polyimides. Several different polyimides have been used as materials in battery separators. A summary will now be provided of those studies.

A series of spongy-like polyimide membranes with different porosities were obtained and characterized in Wang et al. [10]. The membranes were prepared by a wet phase inversion process is on the use of a soluble polyimide. They exhibited excellent thermal stability and dimensional stability with a glass transition temperature of 274°C and thermal shrinkage values of less than 1% at 200°C.

Furthermore, separators for lithium-ion batteries were prepared by phase inversion with a soluble polyimide containing a trifluoromethyl substituent [11]. The membranes have a sponge-like structure with abundant interconnected pores and a porosity of 67%. The membranes also maintain their intrinsic dimensions even after heating at 200°C.

In addition, porous polyimide separators were prepared by using polyamic acid precursors. The production method was non—solvent induced phase separation and a thermal imidization process. The separators that were produced had high thermal stability.

Finally, a novel polyimide battery separator was successfully prepared by non—solvent induced phase separation with two porogens, dibutyl phthalate and glycerin [12]. The separators showed low shrinkage at elevated temperatures. Also, compared with commercial PE separators, the polyimide separator exhibited significant thermal stability.

Along with examining polymers other than polyolefins in battery separators for wet process technology, two approaches are being examined in dry process technology to improve the high-temperature performance of the separators. First, inorganic fillers such as calcium carbonate and titanium dioxide are being added to polyolefins. The filler materials are added to the polyolefin polymer during the extrusion process itself. This is much different from applying a ceramic coating to the separator after it is produced. Because this approach does not add a step in the separator production process, the cost to produce the separator will be less than for ceramic-coated separators.

The concentration level of filler that can be effectively used to produce separators is about 30 wt.%. Above that filler level, it is difficult to achieve the appropriate structure and morphology to produce pores in the separator. This is likely because of the increase in melt viscosity observed with the addition of filler to the polymer. Some tests were performed on batteries made using separators that contain inorganic fillers. Nail penetration tests showed some improvement compared with results that were obtained in batteries containing separators without inorganic fillers present.

A patent was issued to Celgard using this technology [13]. In that patent, the improvement in battery performance during abuse testing, such as nail penetration tests, is attributed to an improvement in the Z-direction dimensional stability of the separator with an increase in temperature. Z-direction refers to the thickness of the separator. Changes in the thickness direction of the separator may be induced by winding the separator during battery fabrication. The patent uses TMA compression data to show the effects of adding inorganic fillers on the Z-dimensional stability of the separators.

Th other area in which additional work has been done on the dry process to enhance the thermal stability of battery separators involved the possible use of polymers other than PE or PP in dry process technology. The original patent that described the dry process battery separator approach discussed the use of the technology for many different polymers other than PE and PP [14]. In particular, high-temperature polymers such as polyamides and polyesters are discussed as being useful in the dry process.

Interestingly, however, not much work has appeared in the open literature about using these other polymers, particularly high-temperature polymers. The one example that has appeared is from Johnson and Wilkes [15]. In the first part of the study, various melt extrusion conditions for poly(4-methyl-1-pentene) (PMP) were presented. The type of stacked lamellar morphology that was obtained during extrusion was found to depend on the resin features as well as extrusion conditions.

The second part of that study [16] investigated the three-stage process used to produce microporous films from the PMP polymer. It was discovered that the morphology of the melt-extruded precursor was critical to controlling the separator's permeability. In addition, the annealing conditions were critical for producing highly microporous PMP films upon stretching. Some limited battery tests performed on cells containing the PMP separators showed improvement in the battery during abuse testing.

The work appears not to have seen follow-up in terms of commercial development. The reasons for that are not obvious. Also, no additional work has appeared in the open literature to determine whether it is possible to produce microporous films from other high-temperature polymers such as polyamides or polyesters. This is an area that requires more attention and study. It is likely that the processing parameters will need to be significantly modified for these other polymers, but work should be done to optimize both the resin and the production parameters to determine how much of an improvement in the thermal stability of the separator can be realized by using these high-temperature polymers.

The next area in which work continues to produce novel high-temperature battery separators is new processes. Some work that has been done on the topic of electrospinning in that regard was presented earlier in this book. This section of this chapter will provide a discussion of new processes being developed to produce high-temperature battery separators.

An aramid membrane with superior thermal resistance was fabricated through a facile papermaking process [17]. Cells made using the separator showed good cycling stability even at an elevated temperature of 120°C. These battery separator features as well as the papermaking process to produce it make the separator a potential candidate for use in high-energy lithium-ion batteries.

This is an example of adopting technologies from other fields to produce lithium-ion battery separators. That a papermaking process is used suggests that the economics for the battery separator will be favorable. Additional work needs to be done on these type of processes to define what polymers can be used and what advantages are offered in terms of battery performance. In addition, the actual translation of process economics into a cost and production model needs to be implemented.

The final area of the continuing development of high-temperature battery separators is in the coating process. An entire chapter of this book is devoted to developments that have occurred using ceramic-coated battery separators. However, another class of materials possesses high-temperature features that can be used as coatings for battery separators. This is the general class of high-temperature polymers. For definition purposes here, a high-temperature polymer is defined as one that has a continuous service temperature higher than 150°C.

Celgard obtained a patent that uses this technology [18]. The patent describes high-temperature separators with a coating layer composed of a polymer that has a melt temperature greater than 160°C. One polymers being particularly useful in the invention is poly(benzimidazole) (PBI). PBI was applied to a polyolefin membrane using a doctor blade as well as a reverse gravure coating method. The PBI polymer was used in a coating solution diluted with dimethylacetamide (DMAc).

Teijin provides a separator coated with their Teijinconex metaaramid [19]. The separator is offered under the trade name LIELSORT. The technology that produces the separator involves simultaneously coating both sides of a base separator. In addition, the high-speed coating technology is five times faster than conventional coating processes.

Fig. 8.2 contains a scanning electron microscopy image of the coated separator. The porosity of the coated layer can be clearly seen. The LIELSORT separator achieves superior heat resistance. In fact, the coated separator maintains its shape up to 250°C. In addition, in spot heating tests, the separator does not break down even at 400°C.

The primary message of these studies of high-temperature separators is that there is more than one way to achieve thermal stability in the battery separator. There is the approach of modifying existing polyolefin separators by incorporating a high-temperature material. That high-temperature material can be an inorganic material or a high-temperature polymer. In that general approach, it is expected that the adhesion between the high-temperature polymer layer and the substrate will not be an issue because both materials are organic polymers. On the other hand, when inorganic materials are used, the presence of a binder polymer is required to guarantee that good adhesion exists between the dissimilar inorganic and organic layers in the battery separators. That is necessary to make sure there is no detachment of inorganic particles from the separator substrate during battery fabrication.

High-porosity separators

The major motivation for producing high-porosity separators is high-power and high-energy batteries used in electric vehicles. Higher porosity in the battery separator means the potential for the absorption of additional electrolyte. There is the possibility for higher-power and higher-energy batteries.

Typical commercial separators have porosities in the 40%–50% range. Mechanisms that primarily control the level of porosity are different in dry process separators from those in wet process separators. In dry process separators, the level of crystallinity of the polymer is a primary factor in

Figure 8.2 Scanning electron microscopy image of LIELSORT-coated separator.

establishing the porosity level possible in the battery separator. For wet process separators, the level of processing oil used in the formulations is largely responsible for establishing the highest porosity value possible. This is because the extraction of oil in the wet process separator is responsible for developing pores.

US Patent 10333125 [20] obtains a high-porosity film using the dry process approach. This is achieved by using a high-molecular weight, low-melt flow index PP polymer resin in the separator. When the resin is melt extruded, it produces a high level of crystalline lamellae content in the precursor membrane. In the dry process, the mode of pore formation in the stretching step involves machine direction stretching. This has the effect of pulling apart stacked crystalline lamellae plates, extending polymer fibrils, and forming rectangular pores. For a dry process stretched membrane, the amount of crystallinity in the lamellae structure in the nonporous PP layer of the precursor membrane is likely an important factor in the formation of the internal porous microstructure. The high-molecular weight, low-melt flow index PP provides a level of crystallinity to the extruded precursor, which may be why the final separator has higher porosity values.

In addition to the higher porosity values that are obtained, improved puncture strength values are reported at these higher porosity levels. This is important because the separator needs to have significant mechanical strength to endure battery assembly as well as the various charge and discharge cycles that occur during the lifetime of a battery. All of these factors contribute to a high level of performance of the battery separator in terms of both the battery cycle life and safety. A high charge rate can be obtained for optimal high-energy performance in applications such as the electric vehicle drive industry.

In the case of wet process separators, the primary mechanism for forming pores is the extraction of processing oil using solvents such as hexane. Because of this, the level of porosity that can be achieved in these separators is directly related to the amount of oil used in the initial separator formulation. However, not all of the oil that is used in the formulation gets incorporated into the actual film. Some of the oil remains on the film surface during extrusion in the production process. This excess oil is eventually deposited on the film production equipment or simply drips off the film. Owing to this, the actual level of oil that is actually incorporated into the film will be less than the oil concentration level used in the extruded formulation. Thus, an upper limit is established for the porosity value that can be established in wet process separators.

The same concern exists with high-porosity wet process separators as was stated for high-porosity dry process separators. The overall concern is the relatively low mechanical strength of separators that are high porosity in nature. However, at the same thickness, wet process separators have higher puncture values than do dry process separators. This is generally because wet process separators are biaxially oriented whereas dry process separators are uniaxially oriented. There is not as large an issue with a decrease in puncture strength with a corresponding decrease in separator thickness. However, if this becomes a concern with high-porosity separators, that issue can be dealt with by varying the molecular weight of the PE polymer used in the formulation. Those polymers present issues in terms of film stretching and other process steps; those concerns need to be addressed before the polymers can be effectively used in wet process battery separator production.

Both approaches to dealing with high-porosity separators represent examples of understanding pore-forming mechanisms in separator production and how those mechanisms can be adjusted as needed by appropriately selecting materials. Nevertheless, using new materials likely also means that the processing parameters for separator production will need to be adjusted as well. This is yet another example of the overall need to have a complete understanding of the structure—property—processing relations of polymers used in battery separators. This is necessary to optimize the use of new methods in emerging applications for lithium-ion batteries.

Electrolyte-absorbing separators

Inferior compatibility between nonpolar polyolefin separators and polar electrolytes results in poor electrolyte wettability. This effect increases the internal resistance of the battery and restricts rate performance [21]. Therefore, the development of an advance separator with good wettability features as well as improved thermal stability is a continuing requirement for lithium-ion batteries.

A great deal of effort has been devoted to overcoming these shortcomings of polyolefin separators. Many of those studies were summarized in Chapter 6 of this book. The use of ceramic coatings with the polyolefin separators is an approach to improving the wettability of battery separators. This section of the current chapter provides a discussion of other approaches developed to improve the wettability features of battery separators.

The surface of polyolefins has been modified through a surface grafting process [22]. That work involved the covalent cross-linking of poly(ethylene oxide) (PEO) onto the surface of a PP separator. The approach used hyperthermal hydrogen cross-linking (HHIC) technology to increase the electrolyte affinity of a PP separator. With HHIC treatment, the polar functionalities of PEO (e.g., −OH, C−O−C) can be preserved through the selective cleavage of C−H bonds and the subsequent cross-linking of the resulting carbon radicals generated on the PEO and PP polymer chains. A PEO coating was found to be firmly fixed on the separator surface; this resulted in significantly improved wetting with the electrolyte.

Another way to improve the electrolyte wettability of polyolefin separators is to use oxygen plasma treatment to surface modify the separators. For example, the surface of a PE separator was modified with functional groups by applying an oxygen plasma [23]. The electrolyte wettability and retention of the PE separators were greatly improved using this approach.

Furthermore, a photo-induced functionalization process was developed to improve the electrolyte affinity of a PP separator significantly [24]. The electrolyte-saturated functional separator showed improved ionic conductivity, an increased lithium-ion transfer number, and thus excellent charge−discharge performance. The results showed that photo-induced functionalization could be a general strategy to provide polyolefin separators with specific properties required for different battery uses.

The wettability of the separators that has been discussed thus far is primarily related to the use of standard electrolyte materials such as carbonates. For certain emerging applications, there is also interest in using other electrolyte materials in lithium-ion batteries. An example of such materials is ILs. The properties of IL electrolytes compared with conventional electrolytes make them more thermally stable, but they also have poor wetting qualities with commercial polyolefin separators.

The development of nonflammable liquid electrolytes based on functional perfluoropolyethers (PFPEs) has opened the door to safe lithium-ion batteries [25]. PFPE is a chemically resistant, noncrystalline that is nonflammable. It also has a low glass transition temperature and low toxicity. It was demonstrated in Huie et al. [26] that a methyl carbonate-terminated PFPE is suitable for use in lithium-ion batteries. However, optimization of the nature of the best separator to use with this new IL electrolyte was necessary to achieve the optimal battery design. Three commercial separators were examined: a PP monolayer, a trilayer PP-PE-PP separator, and a PTFE separator. The ionic conductivities of the PFPEs was measured as a function of temperature in the three different separators. Some differences were observed in the results, which implies that there is an optimal separator structure for the use of this type of electrolyte material. More work should be done to understand separator features related to observed differences in ionic conductivities reported in this study.

In another study [26], a systematic investigation was performed of the physical properties of IL-based hybrid electrolytes. The work involved a quantitative characterization of the electrolyte—separator interface through contact angle measurements. An inverse trend between ionic conductivity and wetting properties was observed for a series of IL-based electrolyte materials. The study revealed that the contact angle influenced the performance of cells more significantly than did the conductivity. This conclusion is based on the fact that cells that contain tetrafluoroborate-based electrolytes with higher conductivity but poorer wetting showed significantly decreased performance relative to cells that contain bis(trifluoromethanesulfonyl)imide electrolytes with lower conductivity but improved wetting properties. This work contributes to the development of new IL-based electrolytes for use in lithium-ion battery systems.

A study characterized separators that use hydrophobic IL electrolytes [27]. Specifically, ILs that are made of N-butyl-N-methyl pyrrolidinium or N-methoxyethyl-N-methylpyridinium were prepared with either bis(trifluoromethanesulfonyl)imide or bis(fluorosulfonyl)imide anions. These materials were tested in combination with several different separators with different chemistries and morphologies. Testing included Gurley and MacMullin numbers as well as a characterization of the separator—electrolyte interface properties.

The influence of the separator on the conductivities of the electrolyte—separator combination was reflected through its MacMullin number. Commercial separators that exhibited a safety feature, either shutdown or zero shrinkage, showed the highest MacMullin number values. In addition, nonfunctionalized microporous polyolefin separators were poorly wetted by some electrolyte materials. The impact of the separator morphology was evidenced by the evolution of the solid electrolyte interface (SEI) resistance and by comparing electrodes cycled using a separator with two different morphologies.

A review was published discussing the development of the optimal separator structure to use with IL electrolytes [28]. That review reached the conclusion that wetting issues that occur in IL electrolyte cells are most likely caused by poor separator compatibility, not electrode compatibility.

A compatible separator needs to be developed before IL electrolytes can be used in commercial lithium-ion batteries. Separators with different processing methods, polymers, additives, and different IL electrolytes were considered in the review. A the main conclusion was that separator studies showed a strong correlation between ionic conductivity and membrane porosity, even more so than electrolyte type. Furthermore, the challenge of a suitable separator for IL electrolytes still had not been solved. Finally, a separator for IL electrolytes will most likely require the combination of a high thermal and mechanical polymer, ceramic additives, and an optimal manufacturing process.

Separator wettability is vital to the electrochemical performance of lithium-ion batteries. The wettability concern becomes more critical for large-capacity electrodes such as those used in electric vehicle applications. In those types of electrodes, the entrance area is limited and the transport distance increases for the electrolyte. The electrolyte wetting process will generally take longer than in smaller consumer electronics batteries.

An unwetted active material will result in underuse of the electrode capacity. Also, there will be an increase in battery resistance that will lead to poor performance of the battery during operation. Furthermore, the electrolyte filling step is often the rate-limiting step in battery fabrication. Hence, any work that can be done to increase the speed of that process will lead to higher production rates for the batteries. Such improvements will certainly have a positive impact on the economics of batteries for electric vehicle applications. This could be a major step forward in further implementing lithium-ion battery technology for these types of uses.

Conclusions

This chapter focused on the unmet needs of the lithium-ion battery separator market. Many of those needs are based on new and emerging applications in areas such as electric vehicles and power grids. The development of these uses is primarily based on the realization that a single separator structure is not the best one for all applications. Thus, the separator structure that is best for uses in consumer electronics devices will likely not be the optimal one for electric vehicle applications. This is primarily because of the larger batteries used in these emerging applications compared with the small batteries used in consumer electronics. Larger batteries are more prone to experience thermal runaway effects. To prevent the occurrence of these phenomena, the separator needs to possess thermal stability at elevated temperatures. This need has not been completely satisfied; work has continued to find alternative solutions to the problem. However, it may be that different size batteries can have different optimal separator structures.

To realize that the best structure has been developed for a specific application, additional work needs to be done to understand the relations between the separator structure and observed battery performance provided by the separator. Some limited efforts have been performed in that area for using IL electrolytes in lithium-ion batteries, but to implement the use of those materials fully, the optimal separator structure needs to be defined. The use of the IL electrolytes could further assist the development of high-temperature separators because they do not experience the same thermal stability issues as do more conventional electrolytes, such as carbonates. Perhaps the solution to developing high-performance lithium-ion batteries with acceptable thermal stability is to incorporate a high-temperature separator as well as an IL electrolyte material.

Other unmet needs discussed in this chapter are largely based on electric vehicle market needs and demands. From a quantitative perspective, the needs of that market will likely depend on the specific size and chemistry of the battery involved in the vehicle. Different battery chemistries may in fact have different optimal separator structures, and this point, they have not yet been completely recognized and appreciated. Specifically, the development of novel electrode and electrolyte materials continues to drive the need for new battery separators.

This brings up one final point about the use of lithium-ion batteries in emerging applications. The battery needs to be considered in its entirety, not as individual components. Thus, as new electrode and electrolyte materials are being developed, it is necessary to realize that the optimal separator may involve adjusting known separator structures. A good example of this has been demonstrated in the case of IL electrolytes, for which the most effective separator structure has yet to be defined. That definition will likely need to be determined to implement the use of IL electrolytes fully in lithium-ion batteries. This is to expected to continue to be the situation as the development of new materials continues for improved lithium-ion batteries. The new way of thinking will certainly lead to new battery structures with have novel features. Those features will likely lead to further developments in new and emerging applications.

References

[1] www.entek.com/lithium - ion.
[2] V. Deimede, C. Elmasides, Energy Technol. 3 (5) (2015).
[3] celanese.con/en/engineered - materials/products/GUR - UHMWPE/product - data.
[4] U.S. Patent 10,347,951, Assigned to Celgard, LLC, July 9, 2019.
[5] H. Yoneda, Y. Nishimura, Y. Doi, M. Fukuda, M. Kohno, Polym. J. 42 (2010) 425.
[6] cmu.edu/ceic/assets/docs/seminar - files/2009−2010/brant - cmu - talk.pdf.
[7] R. l'Abee, F. DaRosa, M.I. Armstrong, M.M. Hantel, D. Mourzagh, J. Power Sources 345 (2017) 202.
[8] Y. Pan, S. Chou, H.K. Liu, S.X. Dou, Nat. Sci. Rev. 4 (6) (2017) 917.
[9] D. Li, D. Shi, K. Feng, H. Zhang, J. Membr. Sci. 530 (2015) 125.
[10] H. Wang, T. Wang, S. Yang, L. Fan, Polymer 23 (2013) 6339.
[11] J. Tan, L. Kong, Z. Qiu, Y. Yan, J. Solid State Electrochem. 22 (2018) 3363.
[12] M. Li, et al., ACS Appl. Mater. Interfaces 12 (3) (2020) 3610.
[13] U. S. Patent 9,570,727, Assigned to Celgard, LLC, February 14, 2017.
[14] U.S. Patent 3,801,404, Assigned to Celanese Co.
[15] M.B. Johnson and G.L. Wilkes, J. Appl. Polym. Sci., 84, 2095 (2002).
[16] M.B. Johnson, G.L. Wilkes, J. Appl. Polym. Sci. 84 (2002) 1076.
[17] J. Zhang, et al., Solid State Ionics 245−246 (1) (2013) 49.
[18] U.S. Patent 10,720,624, Assigned to Celgard, LLC, July 21, 2020.
[19] teijin.com/rd/technology/separator/.
[20] U.S. Patent 10,333,125, Assigned to Celgard, LLC, June 25, 2019.
[21] K. Gao, X. Hu, C. Dai, T. Yi, Mater. Sci. Eng. 131 (2006) 100.
[22] C. Man, et al., J. Mater. Chem. A 2 (2014) 11980.
[23] S.Y. Jin, et al., J. Ind. Eng. Chem. 45 (2017) 15.
[24] W. Zhu, Z. Wang, Q. Weijian, Y. Xiang, J. Electrochem. Soc. 165 (9) (2018) A1909.
[25] D.H. Wong, et al., Chem. Mater. 27 (2015) 597.
[26] M.W. Huie, R. Dileo, A. Marschilok, K.J. Takeuchi, ACS Appl. Mater. Interfaces 7 (22) (2015).
[27] M. Kirchhofer, E. Pallard, S. Passerini, Int. J. Mol. Sci. 15 (8) (2014) 14868.
[28] C.F. Francis, I.L. Kyratzas, A.S. Best, Adv. Mater. 32 (18) (2020).

Further reading

[1] H. Zhang, C.E. Lin, M.Y. Zhou, A.E. John, B.K. Zhu, Electrochim. Acta 187 (2016) 125.

CHAPTER 9

Future developments in battery separators

Introduction

Lithium-ion batteries were initially developed for use in consumer electronics devices. The separators used in those applications are primarily polyolefins, either polypropylene (PP) or polyethylene (PE). The separators are produced using either a wet or dry process approach. Both technologies have been developed extensively and many details have worked out well.

As new applications have emerged for lithium-ion batteries, adjustments and modifications have been made to the basic dry and wet process separator structures. For example, to obtain separators for use in electric vehicle (EV) applications, ceramic-coated separators were developed. Such separators have proven to be effective in providing increased thermal stability and wettability features. However, actual translation of those separator features into improved battery performance has not been completely quantified with different ceramic coatings that provide various levels of enhancements in battery behavior.

These developments are an example of defining separators for a specific need. However, those modifications may need to be adjusted and fine-tuned depending on the battery size and battery chemistry being used in the specific device or vehicle. Those specifications will dictate the optimum separator for use in the battery. It cannot be assumed that the best separator for use with one type of battery will necessary be the best one for use with a different chemistry. Each chemistry must be evaluated as a separate quantity, and the entire battery must be evaluated in its entirety, not as individual components.

This is a relatively unique way to treat lithium-ion batteries and the materials used in their fabrication. In the past, individual components were often examined and the effect of replacing one material in the battery on the effects on the other materials was not completely considered. The entire battery system was not optimized as a total package. Thus, the best performance of the lithium-ion battery was not realized.

Polymer-Based Separators for Lithium-Ion Batteries
ISBN 978-0-12-820120-6
https://doi.org/10.1016/B978-0-12-820120-6.00006-4

This concept is particularly important because new electrode and electrolyte materials are being developed and used. Different electrode materials mean that the overall battery chemistry will change and that the optimum separator may not be used. This is because there may be undesirable chemical reactions with the use of new electrodes that the separator may need to prevent. Depending on the specific battery chemistry being used, the separator will need to be designed according to the desired requirements. Each particular battery chemistry may have an optimum separator structure for the best battery performance to be realized.

This chapter will further expand on that general idea and how it will lead to future developments in battery separators. Those developments will largely be driven by the emergence of new applications in lithium-ion batteries, such as occurred in the EV market. The need to adjust current separator structures or investigate novel production processes will be discussed, and the pros and cons of pursuing both approaches will be highlighted.

Consumer electronics

Whereas many developments in lithium-ion battery separators have been driven by the EV area, developments are also occurring in the consumer electronics field that could lead to the need for new separator structures. The demand for smaller, lighter batteries is large and growing rapidly. Designers of consumer electronics have a difficult choice between runtime and device features that are imposed by the limited energy density of current lithium-ion cells.

The need for higher-density lithium-ion batteries will require the separator structure to be modified. This is because the higher-energy batteries will likely mean that more heat will be generated during battery operation. This increases the potential for thermal runaway effects in these types of batteries. These batteries are smaller than in EV applications, but there is the possibility of fire in devices that use the batteries.

Ceramic-coated separators primarily been developed for EVs may be useful for these consumer electronics applications as well. However, it is not clear that the same formulation is the optimal for the two different applications. This is because each has unique features and there is no one single separator structure that is best for all potential uses. Certainly, the economics of the separator is an important factor for its use in developing consumer electronics. Thus, less expensive approaches for ceramic coatings need to be investigated further. Nevertheless, the safety that separator provides to the battery cannot be compromised for the sake of lower-cost products.

Another development occurring in consumer electronics is faster devices. Consumers want to have electronic devices that respond quickly. The battery separator needs to be able to respond quickly to these changes. Adjustments to the structure of current separators will need to be made to deal with those developments. Wet process separators are a focus of consumer electronics because they can be more easily produced at lower thicknesses than can dry process separators. This is because wet process separators generally have a higher puncture strength values than do dry process separators of the same thickness. The wet process separators can be down-gauged without as large as a decrease in puncture strength and with the corresponding safety that the separator provides to the battery.

Functional separators are being considered for use in next-generation high-energy lithium-ion batteries [1]. They are important developments because they are stable, safe, smart, and sustainable. Functional separators with selective permeability can be applied to retard unwanted migration of specific species to achieve stable cycling operation of the battery. For example [2], an acid-scavenging separator was developed to prevent the effect of acidic species such as hydrogen fluoride (HF), which can promote various parasitic chemical reactions. Among those reactions, the dissolution of transition metal ions and the loss of electroactive as well as transport lithium ions have the most detrimental consequences for lithium-ion battery performance. Cells based on functional separators retain more capacity than do cells with plain PP separators. In addition, cells with functional separators have half the interfacial impedance of those with PP separators. It is speculated in that work that benefits provided by acid-scavenging separators can extend to a broad range of different cell chemistries.

Chemically active separators that use the concept of trapping transition metals, especially Mn, were exemplified by polymers functionalized with 18-crown-6 and 15-crown-5 [3,4]. However, although work with polymeric crown ethers showed the overall technical soundness of the trapping separator concept, crown ethers are expensive in practical batteries. This shifted the general focus of the work to commercial ion-exchange resins and compounds, or simple modifications of these materials.

Multifunctional chemically active separators can result in improved rate performance as well as capacity use and retention in cells with various cell chemistries. The observed performance improvements in the batteries are the result of reduced impedance of the electrodes. This can be related to the presence of thinner and improved electrode surface films. Furthermore, this is attributed to lessened parasitic reactions that occur owing to the use of functionalized separators in the batteries. Those reactions involve the decomposition of $LiPF_6$-based electrolyte solution components. Reduction of those reactions leads to improved cycling performance of cells that contain these functionalized separators.

Another approach to dealing with the dissolution of metal ions was developed [5]. It involves a multifunctional polymeric separator with a lithium-ion—exchanged zeolite coating. The process involves a simplified process to exchange undesirable ions in zeolite with lithium ions. It is proposed that the zeolite coating can provide multiple functions to improve the cycle performance of batteries. First, it absorbs trace amounts of water to reduce the formation of HF. Second, the coated separator acts as a scavenger for HF to reduce its attack on the electrodes. Finally, the zeolite coating acts to trap transition metal ions. All of these effects have the overall impact of having fewer side reactions during battery operation.

A multifunctional polybenzimidazole porous membrane with superior fire resistance, excellent thermal stability, and wettability was prepared [6]. Along with providing excellent thermal stability, the membranes have the ability to regulate the flux of lithium ions owing to the existence of polar groups that exist in the polymer. In addition, the polar groups improve ionic conductivity. Cells that contain the membranes have stable cycling performance at 120°C. The developments can be a new direction for next-generation separators and may lead to a breakthrough for high-performance lithium-ion batteries.

Another way to achieve a separator that has improved thermal stability and leads to the improved transport of lithium ions is to use polyimide as the base separator with a zeolite filler [7]. In this case, the separator has enhanced thermal stability and no shrinkage up to 180°C. The composite separator also has high ionic conductivity. The results show that the composite separator is a promising candidate for use in high-performance and high-safety lithium-ion batteries.

Finally, magnesium borate fiber (MBO) was used as a functional ceramic material to coat a PP separator [8]. The coating layer increases lithium-ion transfer in the $LiPF_6$ electrolyte because the MBO acts as Lewis acid sites that interact with the Lewis base PF_6^-. The increase in the lithium-ion transference number reduces concentration polarization and promotes the migration of lithium ions. Cells made with the separator had improved cycle stability and a higher capacity retention rate at high current levels.

Lithium-ion batteries are applied in a wide range of consumer goods such as smartphones, wearable items, laptops, digital cameras, remote controls, and gaming devices. In many of these applications, the need for lightweight batteries is becoming more important. Lighter materials or an increase in separator porosity are two ways in which the weight of the separator can be reduced in the overall battery weight. It is not clear what level of weight reduction is possible with either of these changes. Also, the relative cost of these changes compared with that of the separator materials needs to be explored. A major question that needs to be addressed is the level of expense consumers are willing to accept for additional convenience in electronic devices. Convenience here covers a wide range of subjects including lighter-weight devices as well as faster devices with a longer-life battery. All of these issues can be partially addressed at least by choosing the appropriate separator structure. However, a single separator structure will not be optimal for addressing all consumer electronics needs. A single basic structure may need to be adjusted and tailored as required. Also, wet process-based separators may be the best choice to satisfy one need whereas dry process separators might be better to address other needs. This is because of the different pore structures observed in the separators, which are produced by the two different production processes [9]. Those different pore structures lead to advantages in different applications; these need to be exploited to employ separator structures effectively in emerging uses in consumer electronics.

Electric vehicles

Separator structures developed for use in consumer electronics likely will not be the optimal structure for use in EVs. A primary reason is the larger batteries used in EVs compared with consumer electronics. This difference in battery size means that there is a larger potential for thermal runaway effects, which translates into a higher possibility for catastrophic events such as fires and explosions.

One of the main ways in which this effect is handled is by using separators that have high-temperature thermal stability. As outlined in a previous chapter, this was achieved through the use of ceramic-coating technology on a base polyolefin separator. The advantages and disadvantages of this approach were discussed earlier in this book.

With technology, the battery may obtain elevated temperatures and the separator will provide safety to the battery by keeping the electrodes separated. It does not necessarily consider separator shutdown and high-temperature stability as providing combined safety features. It is still debated whether shutdown or high-temperature melt stability is the most important safety feature provided by the separator. The answer will likely depend on the size of the battery being used in the specific application.

One approach that has been pursued to deal with this issue is the development of a low-temperature shutdown separator. For example, shutdown can be achieved at 100−110°C through the use of an appropriate coating [10]. The coatings that were used in that work were ultrathin, less than 2 μm thick. This coating process was achieved with no interference to any performance requirements of the separator. The work is primarily based on the use of a polymer latex material using styrene acrylic polymer particles. Additional details of this technology can be found in US Patent 9070954 [11].

A different approach to achieving low-temperature shutdown was used in US Patent 6749961 [12]. That work used a blend of high-density PE polymer and a PE wax. Owing to the presence of the PE wax in the mixture, shutdown temperatures less than 120°C are reported in that patent. This is certainly lower than what was achieved with high-density PE alone, but it is still not in the 100−110°C range.

The use of PE types other than high-density PE offers another way to achieve low-temperature shutdown separators. Polymers such as low-density PE have lower melting points than high-density PE. For example, the melting point for average commercial low-density PE is typically 105−115°C. This is certainly within the range of the stated low-temperature shutdown that is desired. However, technical issues are involved in implementing low-density PE into a battery separator. First, it is not easy to produce pores in this polymer. It does not appear to have the appropriate crystallization features for use in dry process technology. An appropriate solvent or plasticizer would need to be found and employed to use low-density PE in the wet process.

However, another factor needs to be considered with the use of low-density PE in a battery separator structure. Because of its relatively low melting temperature of about 105−115°C, it typically will be processed in the melt at 170–200°C. This is significantly lower than the processing temperature used with a high-temperature polymer or even PP. There is a significant mismatch in the processing temperatures for the two different classes of polymers used in the separator structure. This problem needs to be overcome with the production of a separator that contains both a low-melting polymer to ensure low-temperature shutdown and a high-temperature polymer to provide the separator with high-temperature thermal stability.

A way to address that concern is to use individual films for the two different polymer types and produce a laminated separator for each. This could be done by producing porous structures in both film components and then laminating the two porous films together. To use this approach successfully, it is necessary to soften or partially melt one of the films to achieve adhesion in the melt lamination process. There cannot be a significant pore closure during lamination. Depending on the nature of the polymers being used, the acceptable temperature processing window may be small. If this is the case, care must be taken to ensure that temperature control guidelines are satisfied.

There are other ways to produce a laminated structure in the separator other than through melting. For example, adhesive lamination offers an alternative to lamination that is performed through a heat and pressure process. However, the chemistry of the adhesive that is used needs to be compatible with the battery chemistry. In addition, the adhesive that is used needs to be thermally stable over the temperature range of normal battery operation.

This also pertains to polymer binders used to adhere ceramic coatings to base polyolefin separator substrates. Polyvinylidene fluoride (PVDF) is one of the polymers used in that application. However, PVDF is limited in terms of the adhesion that it can provide at elevated temperatures because it melts at about 165°C. To deal with this relatively low melting point, binder polymers with increased thermal stability need additional exploration.

Some work has been done in that area. However, little effort has been made to understand the effect of high melting point binders on the thermal behavior of composite separators as well as the electrochemical performance of cells that use those separators. One exception to that is the work presented in Yu et al. [13]. In that research, a case study is presented of a high–melting point poly(vinyl alcohol) (PVA) for a ceramic coating of aluminum oxide on polyolefin separators. The effect of different contents of PVA was the primary focus of the work. The use of PVA improves the thermal stability of the separator compared with PVDF. The research concludes that to have high-performance composite separators, it is necessary to use a binder with a high melting point, good electrolyte wettability, and strong bonding with the coating particles. The binder content should be well-controlled. The coating layer needs to be constructed with low tortuosity to be greatly beneficial for cell performance. The study provides insights into the design of high-performance composite separators with improved safety features.

Another approach is to use a thermally stable polyimide binder with an aluminum oxide coating [14]. Ultrafine aluminum oxide and polyimide were coated onto the surface of a porous PE separator. The coated separators showed no thermal shrinkage up to 160°C. These results showed that the separator had high thermal stability. The prepared separator is a promising candidate to improve the thermal stability and capacity retention of lithium-ion high power batteries. Additional work should optimize the structure of the polyimide binder in terms of the battery performance that can be achieved.

There has also been patent activity in the development of high-temperature binders for use with ceramic coating materials. US Patent Application 20190355953 [15] uses a mixture of materials including tannic acid and xanthan gum. Specifically, the patent involves a binder that contains a large amount of —O and hydroxyl groups in the molecule. The binder exhibits thermal stability as well as an enhancement in adhesion, adsorption of a transition metal, and self-healing. Ideally, the binder should contain 10 wt% or more of a hydroxyl group per molecule to be most effective for use in the battery separator.

Additional work needs to be done to understand the chemical structure of a binder polymer that is the most effective in the ceramic coating formulation. There may be a most effective chemical group to improve adhesion between the polymer and inorganic material. However, that functionality will likely be different for different polymer—inorganic combinations, and thus needs to be realized as different systems are being studied and developed.

In addition to the development of more effective binder materials, more quantitative methods need to be developed to define the effectiveness of the binders. One currently used test procedure is a peel strength test, which measures the adhesion of the ceramic coating to the base substrate. It is a measure of the interaction between the coating and the base substrate, in which higher peel strength values indicate stronger interactions.

However, there are several issues with the peel testing that is typically done. The performance of the coating system is never solely the result of the material properties. Often, the test results are a function of the test that is being used rather than a function of the material that is being tested. In fact, different peel tests have been shown to yield different results. Specifically, 90 and 180°C peel tests have resulted in different adhesion strengths for the same ceramic-coated separator [16]. These results indicate that direct comparison of the adhesive strengths of different formulations requires which test procedure, 90 or 180° peel test is being used. Furthermore, even within a specific test, the test parameters need to be specified because it is likely that the measured peel strength results will depend on the actual test parameters being used.

A test that simulates the conditions of the separator under normal battery fabrication and operation needs to be agreed upon. Different peel tests can yield quantitatively different results. Thus, there is a need for the development of a standard test and associated test parameters that can be used to compare different formulations and different separator structures. The best test in terms of providing information related to specific performance in batteries is one that mimics the situation for the separator in the battery environment. Particularly significant in that regard is the strain rate of the peel test used to simulate the battery fabrication process adequately.

In addition, the effect of temperature on adhesion features needs to be further explored. Inorganic particles should not delaminate from the separator substrate at elevated temperatures. If this occurs, there inorganic particles might be removed from the separator, which might cause issues and the potential for particles to cause holes in the separator.

Another development in separators that directly affects the EV market is high-porosity separators. Commercial separators have porosity values in the 40%–45% range. This porosity limits the amount of electrolyte that can be absorbed by the separator. This will affect the energy and power density that can be achieved with the battery that contains the separator. Thus, to achieve high-performance batteries, one approach is to use a separator with relatively high porosity.

There have been several attempts to produce high-porosity separators. For example, cellulose-based composite nonwovens have been used to fabricate a separator with 70% porosity [17]. Compared with a PP separator with 55% porosity, the composite separator exhibited improved electrolyte uptake. Cells that use the composite separator also exhibited better rate capability and cycling retention than cells that contained the PP separator.

This is because of facile ion transport as well as excellent interfacial compatibility. In addition, cells with the composite separator provided stable cycling performance at an elevated temperature of 120°C.

In addition, a flexible SiO_2 nanofiber membrane combined with a PVDF-HFP (hexafluoropropylene) nanofiber membrane was prepared by electrospinning [18]. The porosity of the composite nanofiber membrane was 89.7%, which was more than twice that of an ordinary PE separator. The liquid absorption rate of the membrane was much higher than an ordinary PE separator and reached 483%. Furthermore, the cycle and rate performance of batteries prepared with the composite membrane improved significantly. These results make the composite membrane a viable alternative for high-power battery separators.

Electrospinning offers the potential to produce battery separators with high porosity levels. However, owing to the high porosity, they also have poor mechanical strength, which is a factor that has hindered their industrial application. An approach to addressing this issue was taken with a polyethylene terephthalate/PVDF separator and involved hot pressing the separator. Use of the hot pressing allowed for the production of a separator with excellent mechanical properties.

To use these high-porosity separators fully, it may be necessary to combine different processing techniques, as was done in the latter work. The electrospinning approach will be useful for producing the initial separator structure with high porosity. However, such a structure will have low mechanical strength owing to the high level of porosity. This could have negative implications for handling these separators.

This issue can be partially addressed by using polymers in the separators that have high mechanical properties. However, these polymers can also have processing issues directly related to their high mechanical strength. This is a usual trade-off that exists between the processing and properties of high-performance polymers. A typical example of this is ultrahigh-molecular weight PE, which becomes harder to orient as the molecular weight increases [19]. However, these types of polymers need to be further investigated because of the high mechanical strength they can provide. They will be particularly useful for separators with high porosity levels.

There has also been some patent activity to produce high-porosity separators. Separators with up to 65% porosity can be produced using a high-molecular weight, low-melt flow index PP resin in the dry process [20]. Use of this PP resin in separator production leads to increased mechanical strength that produces a balance with the higher porosity levels in the final separator. That work demonstrated that a change in the melt flow index of the resin can have a significant effect on crystallization behavior that is important for developing the pore structure using the dry process technology. Additional investigations using this approach are warranted to determine what level of balance between the separator porosity and its mechanical properties can be achieved. The patent says that separators with high porosity will have uses in EV applications.

The advent of EVs has led to the need to reexamine the structure of separators used in lithium-ion batteries in consumer electronics. The conclusion was that separators used in one application are not necessarily the best ones for use in other applications. This is primarily because of the larger size of the batteries used in EVs. That size difference means there will be different safety requirements for these batteries compared with the smaller batteries are used in consumer electronic devices.

In addition to the safety requirements that are part of EV batteries, higher-energy batteries are a requirement for these developments. One way that goal is being achieved is by using high-porosity separators with porosity levels higher than standard separators. This need may lead to unique processing scenarios such as electrospinning to produce separators. It may be that using these unique processing schemes will allow for the implementation of new materials into battery separators. This combination of novel materials and processes will lead to a new generation of battery separators for use in lithium-ion batteries.

Rising environmental concerns and increased greenhouse gas emissions have compelled several countries to rethink the use of fossil fuels in the transportation sector. Electric vehicles are a possible replacement for fossil-powered vehicles. Thus, several countries are introducing plans to shift the automotive industry from diesel and gas vehicles to EVs. Increasing sales of EVs will drive the demand for lithium-ion batteries, which will significantly contribute to the growth of the battery separator market. Furthermore, those developments will lead to major innovations and developments in new battery separator structures and the use of new materials and processes to produce those structures.

Power grids

Innovations in solar, wind, and other renewable energy sources have become more prominent within the electrical generation system. However, to use the full potential for the needs of the power grids, there is a requirement for bigger, more reliable battery solutions. Heavier demands are placed on these batteries compared with those used for both EVs and mobile devices. This is because of the constant nature of electrical energy charge—discharge cycles, storage, transmission, and distribution.

Batteries to meet these needs will use material innovations, including silicon anodes, to increase charge—discharge cycles and energy density. Use of these new electrode materials means the separator material needs to be reexamined to determine the best structure that can be used with these new electrodes. It is not obvious that current separators will be the optimal ones for these novel batteries. More work needs to be done to establish that whether that is a valid assumption.

Primarily because of the large size of the batteries often used in these power grid applications, many of the same safety requirements that exist in EVs are present in these uses. Thus, the ceramic-coated separators primarily developed for use in EVs will be a good starting place for separators in these batteries for use in power grids. However, it is important for it to be verified that current ceramic-coated separators are optimal for power grids. Otherwise, the formulation used in ceramic-coated separators for EVs may need to be modified to deal with the needs of power grid batteries.

One approach that may be useful involved mixing low-melting point PE microspheres into the ceramic-coating formulation to produce a functionalized PE separator [21]. This lowers the shutdown temperature from 135 to 110°C. Pouch batteries prepared with the novel separator showed improved tolerance to overcharge effects as well as internal short circuits. These effects are achieved without sacrificing the electrochemical performance of the batteries. These results provide important implications for the further development of safer separators for use in lithium-ion batteries.

Developments such as power grids that use lithium-ion batteries will continue to drive the need for new battery materials. As new electrodes and electrolytes become common, an examination of the battery separator being used in the battery is critical to guarantee that the optimum separator structure is used. This is a requirement to be certain that the most efficient use of lithium-ion batteries is occurring in these emerging applications.

Other uses for lithium-ion batteries

Along with three areas that were highlighted and discussed as applications for lithium-ion batteries, another needs to be discussed: high-charge rate batteries for use in power tools. An issue with this type of battery is that the use of high currents will heat and degrade a battery faster than slow trickle charge. This will directly affect the lifetime of the battery, which will likely need to be replaced often.

With the increase in heat generated in these batteries, detrimental chemical reactions can be initiated that further affect the performance of the battery. Separators can be designed to scavenge the by-products of these undesirable reactions to limit their overall effect. That can be accomplished by using functionalized coatings or additives that can eliminate the reaction results. However, these approaches must not negatively affect either the separator mechanical properties or the overall electrochemical performance of the battery that contains the separator.

Another area in which lithium-ion batteries are generating interest is civil and defense applications. Specifically, underwater applications that the primary source of power for submarines and unmanned underwater vehicles are being examined. Lithium-ion batteries offer higher energy density and a long service life compared with their predecessors.

This need will certainly spur additional developments in battery separators. A primary concern of lithium-ion batteries in these uses is the potential for thermal runaway and resultant fires. These concerns can be initially addressed by using ceramic-coated separators. However, it still needs to be determined whether the separator structure and coating formulations that are used will function as the optimal structure in these civil and defense uses.

These applications, such as power tools and defense, will require a reexamination of the separator structure to guarantee that the optimal materials are being used in the most efficient manner. This is a general theme of all of these developments in both power tools and defense applications for lithium-ion batteries. As these new applications continue to emerge, requirements for the separators will change and the separator materials and structure will need to be adjusted as needed.

This chapter has discussed future developments in lithium-ion battery separators. Those advances will largely be driven by new uses for lithium-ion batteries. Those new uses will occur in several different areas including EVs and power grids. Different products being developed in these areas will require a close look at features of the separators that provide advantages to batteries and how those features can be tailored. This is the entire area of structure—property—processing relations for the separators that needs additional study; research and analysis have been inadequate and more work is required.

This idea follows the general theme that each application should be examined in detail in terms of its requirements and how they can best be satisfied by the lithium-ion battery, including the separator. It should not be accepted that if a certain structure works in one use, it is automatically the optimal structure for all applications. Each development has unique features and should be examined on its own.

Finally, the battery needs to be considered in its entirety, not as separate components. For example, for power grids, silicon electrodes are considered to be one battery component. It is not clear whether that choice of electrode material will mean that a different separator and perhaps a different electrolyte should be used. In addition, work is ongoing on novel electrolyte materials; the use of those electrolytes may lead to the development of unique separator materials. Additional work is definitely needed in this area confirm that the most effective battery is being used in these emerging applications. It is only through this optimization that lithium-ion batteries can be exploited to the fullest extent.

References

[1] Y. Pan, S. Chou, H.K. Liu, S.X. Dou, Nat. Sci. Rev. 4 (6) (2017) 917.
[2] A. Bannerjee, B. Ziv, Y. Shulina, S. Luski, D. Aurbach, I.C. Haladay, ACS Energy Lett. 2 (10) (2017) 2388.
[3] Z. Li, A.D. Pauric, G.R. Goward, T.J. Fuller, J.M. Ziegelbauer, M.P. Balagh, I.C. Haladay, J. Power Sources 272 (2014) 1134.
[4] B. Ziu, et al., J. Electrochem. Soc. 161 (2014) A213.
[5] J. Xu, X. Xiao, S. Zeng, M. Cai, M.W. Verbrugge, ACS Appl. Energy Mater. 1 (12) (2018) 7237.
[6] Sun, et al., RSC Adv. 9 (2019) 40084.
[7] Y. Li, et al., Polymers 12 (4) (2020) 764.
[8] X. Wang, et al., Chem. Electrochem. 7 (5) (2020) 1187.
[9] G. Venugopal, J. Moore, J. Howard, S. Pendalwar, J. Power Sources 77 (1) (1999) 34.
[10] www.sbir.gov.sbirsearch/detail/258210.
[11] U.S. Patent 9,070,954, Assigned to Optodot Corporation, June 30, 2012.
[12] U.S. Patent 6,749,961, Assigned to Celgard, Inc, June 15, 2004.
[13] L. Yu, Y. Jin, Y.S. Lin, RSC Adv. 46 (2016).
[14] C. Shi, et al., J. Membr. Sci. 517 (2016) 91.
[15] U.S. Pat. Appl. 20190355953, Assigned to LG Chem, Ltd, Jan. 5, 2018.
[16] H. Jeon, J. Choi, M. Ryou, Y.M. Lee, ACS Omega 2 (5) (2017) 2159.
[17] J. Zhang, et al., Sci. Rev. 4 (2014) 3935.
[18] Y. Xu, J.W. Zhu, J.B. Fang, X. Li, M. Yu, Y.Z. Long, J. Nanomater. (2020), 3879040.
[19] X. Wang, Y. Mu, G. Zhao, J. Gao, Mater. Sci. Forum 993 (2020) 906.
[20] U.S. Patent 10, 333, 125, Assigned to Celgard, LLC, June 25, 2019.
[21] X. Chen, S. Chen, Y. Lin, K. Wu, S. Lu, Ceram. Int. 46 (15) (2020) 24689.

Index

Note: 'Page numbers followed by "*f*" indicate figures and "*t*" indicate tables.'

R

S

FSC
www.fsc.org
MIX
Papier | Fördert gute Waldnutzung
FSC® C083411

Kolibri 360 GmbH
Ferdinand-Jühlke-Straße 7
99095 Erfurt, Deutschland
produktsicherheit@kolibri360.de